Managing Infrastructure Projects

Second Edition

World Scientific Series on the Built Environment

ISSN: 2737-4831

Series Editor: Willie Chee Keong Tan
National University of Singapore, Singapore)

Published

Vol. 7   *Managing Infrastructure Projects*
Second Edition
by Willie Tan

Vol. 6   *Urban Greening Techniques: An Introduction*
by Chun Liang Tan

Vol. 5   *Urban Management: Managing Cities in Uncertain Times*
by Willie Tan

Vol. 4   *Facilities Planning and Design: An Introduction for Facility Planners,*
Facility Project Managers and Facility Managers
Second Edition
by Jonathan K M Lian

Vol. 3   *Managing Infrastructure Projects*
by Willie Tan

Vol. 2   *Introduction to Workplace Safety and Health Management:*
A Systems Thinking Approach
Second Edition
by Yang Miang Goh

Vol. 1   *Contract Administration and Procurement in the Singapore*
Construction Industry
Second Edition
by Pin Lim

World Scientific Series on the Built Environment

Volume 7

Managing Infrastructure Projects

Second Edition

Willie Tan

National University of Singapore, Singapore

World Scientific

NEW JERSEY · LONDON · SINGAPORE · BEIJING · SHANGHAI · HONG KONG · TAIPEI · CHENNAI · TOKYO

Published by

World Scientific Publishing Co. Pte. Ltd.

5 Toh Tuck Link, Singapore 596224

USA office: 27 Warren Street, Suite 401-402, Hackensack, NJ 07601

UK office: 57 Shelton Street, Covent Garden, London WC2H 9HE

British Library Cataloguing-in-Publication Data
A catalogue record for this book is available from the British Library.

World Scientific Series on the Built Environment — Vol. 7
MANAGING INFRASTRUCTURE PROJECTS
Second Edition

Copyright © 2024 by World Scientific Publishing Co. Pte. Ltd.

ISBN 978-981-12-8598-1 (hardcover)
ISBN 978-981-12-8658-2 (paperback)
ISBN 978-981-12-8591-2 (ebook for institutions)
ISBN 978-981-12-8599-8 (ebook for individuals)

For any available supplementary material, please visit
https://www.worldscientific.com/worldscibooks/10.1142/13667#t=suppl

Desk Editors: Gregory Lee/Amanda Yun

Typeset by Stallion Press
Email: enquiries@stallionpress.com

Preface

This book is about practical ways to manage the infrastructure development cycle from project initiation to the end of the operation and maintenance phase. It focuses on the Public–Private Partnership (PPP) contract and, from this perspective, private and public sector procurement are variations. Project management is about managing time, cost, quality, safety, and risks using project management methodologies and risk instruments, as well as regulatory, governance, and contract structures. If we leave these critical activities to intuition, projects are likely to fail.

The idea of writing this book originates from a series of Development Finance lectures in the MSc (Project Management) program at the Department of the Built Environment, College of Design and Engineering, National University of Singapore. The students come from diverse backgrounds, such as information technology, business, architecture, quantity surveying, urban planning, project management, engineering, construction, facilities management, transport, finance, economics, and law. The book provides a structured guide to these diverse students as well as to researchers, public officials, project sponsors, lenders, developers, contractors, subcontractors, suppliers, investors, infrastructure fund managers, insurers, facilities managers, non-government organizations, and project consultants such as designers, engineers, environmental specialists, legal advisors, and brokers.

My approach is to focus on general principles that are practical and applicable in different countries, particularly in the developing world where markets and regulatory institutions are less developed. Hence, the book uses many examples to illustrate applications and avoids detailed case studies. The book deals with only the main processes, leaving it to the reader to explore the details from standard textbooks in project finance and project management.

The infrastructure development cycle is considerably longer than the usual project cycle found in standard textbooks on project management. There are many stakeholders in an infrastructure project. Hence, we may view the management of such projects from different perspectives. This book focuses on the role of the public *grantor* during the initial phases of identification, feasibility study, project preparation, risk allocation, and public tender. It then considers the private *sponsor's* bid preparation, the *lender's* due diligence, and the PPP contract award. Thereafter, we track the pre-construction activities to establish the special purpose vehicle, set up the project governance structure, design the project, and prepare for the construction tender. If it is a Design-Build or Engineering, Procurement, and Construction (EPC) project delivery method, then the *contractor* is also responsible for the design. During the mobilization and project execution phases, the discussion shifts to the relation between the sponsor and contractor. Upon project completion, the sponsor appoints an *operator* to manage the operation and maintenance. Finally, towards the end of the PPP contract, the sponsor hands over the project to the grantor.

Project terminology differs across countries, sectors, and analysts. In this book, the grantor is the ceding authority, public agency, or public entity. The sponsor is the client, owner, developer, or employer, depending on the context. Sometimes, we use the plural form; that is, sponsors instead of sponsor. "Designers" are architects, engineers, and other design professionals. The following terms are similar: bid or tender, variation order or change order, and contract or agreement. It differs from the use of an "agreement" as an unenforceable private understanding or arrangement. I use the pronoun "he" or "she" interchangeably and "$\log x$" instead of "$\ln x$" for natural logarithm to avoid confusing "n" with sample size, as in "$\ln n$." Unless otherwise stated, the monetary unit is Singapore dollars.

I assume readers are familiar with basic project management, economics, finance, calculus, and statistics or data analytics. Where possible, I avoid the heavy mathematics and focus on intuitive understanding of the problems and solutions. I also use many examples on how to put theories into practice.

In this second edition, many chapters have been expanded, resulting in a thicker book. While the first edition focuses on the first half of the infrastructure development cycle, the second edition expands on the topics in the second half of the cycle to provide a more balanced exposition.

I thank the following academics and practitioners for their helpful suggestions on the structure of the book and chapters: Alexander Lin (NUS), Bijay Joseph (Chuan Lim Construction), Imriyas Kamardeen (Deakin University), Calvin Yeung (Changi Airport Group), Christopher Leong (IDA PM), Dilini Thoradeniya (University of Moratuwa), Eoon Hoon Eng (Exyte Singapore), Eugene Seah (Meinhardt Singapore), Jeffrey Tan (Vopak Asia and Middle East), Jiang Hongbin (Asian Development Bank), Jonathan Lian (NUS), Li Dezhi (Southeast University), Lim Pin (NUS), Neo Kim Han (Housing and Development Board), Prashant Anand (IIT Kharagpur), Ravi Shankar (National University Health System), Shen Liyin (Zhejiang University City College), Stephen Tay (NUS), Winston Hauw (NUS), and Zhang Yajian (Amazon).

Lastly, I thank Amanda Yun and Gregory Lee at World Scientific Publishing for their kind assistance and encouragement to bring this book to fruition amid my busy schedule.

Willie Tan

Contents

Preface   v

Chapter 1   Introduction to infrastructure   **1**

Physical infrastructure   1
Soft infrastructure   1
Ownership and performance   2
Infrastructure investment   3
Infrastructure as an asset class   4
Long waves of infrastructure investment   6
Sustainable infrastructure   8
Green infrastructure   9
Resilient infrastructure   11
Stakeholders   12
Complexity   13
Environmental, social, and governance framework   13
Challenges   15
References   17

Chapter 2   The stages of infrastructure development   **19**

Institutional framework   19
The project cycle   19
Phases in infrastructure development   20
Approaches to project management   22
Leadership and teamwork   23
Integration   26
References   26

Chapter 3   Project identification   **29**

Sources of projects   29
Project goals and objectives   30
Preliminary screening   30
Assessment of options   30
Preliminary study   32
Technical assessment   32
Economic appraisal   37
Project benefits   38
Benefit from marketable output   38
Benefit from non-marketable output   40
Benefit from transport cost savings   40
Other benefits   43
Non-benefits   44
Double-counting   45
Project costs   45
Imported inputs   46
Investment costs   47
Operating costs   48
Working capital   49
External costs   50
Statement of benefits and costs   53
Decision criteria   54
Pareto criterion   54
Kaldor–Hicks criterion   54
Net present value   55
Choice of discount rate   56
Internal rate of return   57
Cost-benefit analysis of a proposed road project   59
Risk assessment   60
Sensitivity analysis   61
Monte Carlo simulation   61
Project options   62
Qualitative method   63
Recommendation   67

Criticisms of cost-benefit analysis   67
Political, regulatory, and legal assessment   68
Business case   68
References   69

Chapter 4   Feasibility study   **71**

Project Brief   71
Feasibility study   72
Output and technology   73
Site analysis   73
Site appraisal   74
Project schedule, milestones, and length of concession period   77
Shadow bid model   77
Political feasibility and user affordability   78
Fiscal sustainability   78
Regulatory and legal assessment   79
Socio-environmental impacts and mitigation strategies   79
Public–private partnership contract structure   80
Project governance structure   81
Bankability   81
Possible government support   81
Market sounding   82
Business case   83
References   83

Chapter 5   Project preparation   **85**

Preparation of documents   85
Request for Qualification   85
Draft public–private partnership contract   87
Draft permits and approvals   88
Request for Proposal   88
Instructions to bidders   90
Response Package   90
Land acquisition   90
References   91

Chapter 6   Tender   **93**

Pre-qualifying bidders   93
Request for Proposal   93
Responding to queries   93
Issue of final tender   94
Tender deposit   94
Reference   94

Chapter 7   Sponsor's bid preparation   **95**

Decision to tender   95
Development Agreement   96
Project team   96
Political, regulatory, and legal assessments   96
Design proposal   97
Price proposal   98
Revenue   99
Market study   99
Stock-flow model   99
Purchase contract   102
Call option   102
Put option   103
Option pricing   103
Procurement strategy   107
Lump-sum contract   108
Design-Build contract   108
Construction Management contract   110
Measurement contracts   110
Cost estimates   110
Schedule estimate   112
Quality   112
Construction   113
Inputs   113
Forward contracts   113
Futures contract   115
Operation and maintenance   116

Force majeure events 116
Sources of funds 116
Cost of funds 118
Amortizing loan repayment 120
Balloon loan repayment 122
Currency risks 123
Currency swap 124
Currency fluctuations 124
Inflation risk 126
Interest rate risk 127
Interest rate swap 128
Financial feasibility 128
Income Statement 129
Cash flow statement 130
Equity internal rate of return 132
Balance Sheet 132
Ratio analysis 134
Project insurances 135
Draft shareholders' agreements 138
Draft project agreements 140
Direct Agreement 140
Construction Contract 140
Operation and Maintenance Contract 142
Off-take Agreement 143
Supply Agreement 143
Other Agreements 144
Business case 144
References 145

Chapter 8   Bid evaluation and contract award   147

Price bids 147
English auction 147
Japanese auction 148
Dutch auction 148
Vickrey auction 148

Sealed bid auction   148
Revenue equivalence   149
Winner's curse   149
Over-bidding   149
Design and price bids   150
Best-value bids   150
Issues in bid evaluation   150
Bid opening   151
Award of contract   151
References   151

Chapter 9   Lender's due diligence   **153**

In-principle approval   153
Borrower's background   154
Project information and structuring   154
Expert panel   154
Loan negotiation   155
Financial agreements   157
Loan Agreement   157
Security Agreement   157
Equity Support Agreement   158
Common Terms Agreement   159
Accounts Agreement   160
Inter-Creditor Agreement   161
Finalization of project documents   161
Hedging arrangements   161
Financial close   162
Refinancing   162
Securitization   162
References   164

Chapter 10 Pre-construction activities   **165**

Establishment of the special purpose vehicle   165
Types of businesses   165
Project governance structure   166

Project Brief   167
Appointment of the project manager   167
Requirements   168
Scope   168
Procurement strategy   169
Appointment of the project team   169
Project schedule   169
Site survey, analysis, and engineering studies   170
Programming   170
Design Brief   171
Design management   172
Schematic design   173
Design development   173
Progressive cost and schedule estimates   174
Design reviews   175
Value engineering   175
Constructability review   176
Regulatory design approvals   177
Managing design information   177
Commissioning plan   178
Development of contract documents   179
Tender documents   179
Provisional and prime cost sums   180
Tender process   180
Contractor's decision to bid   180
Contractor's bid estimate   180
Award of construction contract   182
References   183

Chapter 11 Mobilization   **185**

Submittals   185
Schedule of values   185
Baseline plans   186
Site organization and staffing   187
Kick-off meeting   188

Award of subcontracts   189
Procurement   189
References   189

Chapter 12   Construction   **191**

Project monitoring and control   191
Construction schedule   191
Resource management   194
Submittals   197
Budget and cost systems   198
Managing quality   198
Managing safety and health   200
Managing scope   201
Managing stakeholders   203
Commissioning plan   204
Managing risks   204
Managing subcontractors and suppliers   206
Inspection plan   207
Documentation plan   207
Communications plan   208
Progress reports   208
Claims and disputes   210
References   212

Chapter 13   Project close-out   **213**

Close-out activities   213
Start-up and testing   213
Punch lists   214
Certificate of Substantial Completion   214
Final payment   214
Handing over of documents and materials   214
Certificate of Occupancy   215
Preparation for operation and maintenance   216
Release of resources   216
Evaluation of performance   216
References   217

Chapter 14 Operation and maintenance   **219**

   Asset management   219
   Strategic asset management   220
   Operation and maintenance contract   221
   Facilities management contract   221
   Principles of maintenance   222
   Artificial intelligence   225
   Digital twin   226
   Asset enhancement   226
   Realizing the asset value   227
   References   227

Chapter 15 Handing over   **229**

   Early termination   229
   Handing over issues   230
   Public asset management   231
   References   232

Index   233

CHAPTER 1

Introduction to infrastructure

Physical infrastructure

Infrastructure refers to durable physical systems that support economic and social activities of a city or country. The physical infrastructure consists of economic and social infrastructure. The former includes

- utilities and energy (water, electricity, gas, oil, solar, and wind);
- transport networks (road, rail, sea, and air);
- waste management systems; and
- communication and information technology systems (cable, mobile, satellite, and broadband networks).

The social infrastructure consists of facilities such as

- schools and institutions of higher learning;
- prisons, courts, and hospitals;
- sports and recreational facilities; and
- religious buildings.

A more general definition of infrastructure includes residential units, factories, offices, and commercial buildings.

Soft infrastructure

An even broader definition of infrastructure includes the "soft" side, that is, the institutions of society that govern common undertakings within the community. Examples of soft infrastructure include

- public and private organizations;
- laws, regulations, and rules;

- social norms; and
- social or business networks.

Organizations such as firms, clubs, religious bodies, government ministries, hospitals, and schools are institutions because they have their own rules. They are the institutional players (North, 1990).

The laws, regulations, and rules of a society cover property rights, currencies, weight and measures, land use zoning, health and safety, and so on. These rules govern the behavior of individuals.

Social norms are unwritten rules of acceptable social behavior, such as the expectation of punctuality and being courteous. Not all norms increase social welfare; some norms may be discriminatory, such as those relating to marriage, gender, and race. These norms may restrict the choices or wishes of those affected, such as the decision to get married, have children, and so on.

There are also rules that govern behavior in social and business networks. Some of these rules are unwritten; that is, they are norms, especially in social networks. Business networks may have explicit rules. These networks tend to benefit members by sharing ideas, resources, and opportunities (Orru *et al.*, 1996). However, they can also have negative effects. For example, close kinship ties can impose certain unwanted obligations on members (Hyden, 2012) or cut off suppliers that are more competitive.

Despite these downsides, norms and networks are sometimes considered as the "social capital" of society (Banfield, 1958; Putnam, 2001), unlike natural, physical, human, or financial capital. They function like "capital" because of their ability to increase production by reducing the transaction costs of doing business or solving collection action problems (Ostrom, 1990), such as by developing trust (Fukuyama, 1995; Ghate *et al.*, 2008), sharing of business information, monitoring performance, mediating disputes, and punishing cheaters (Coase, 1937).

Ownership and performance

Ownership of physical infrastructure may be private, public, or mixed, such as the mix of public and private power generation plants, roads, schools, universities, hospitals, and sports facilities.

It is sometimes argued that public managers face weaker incentives because they rely on public funding, do not face threats of takeover if performance falters, pursue political and social objectives, and are difficult to monitor for performance (Frydman *et al.*, 1999).

However, it is not clear that private managers act in the interests of shareholders if they own little stock; that is, there is the well-known problem of corporate separation between ownership and control. Further, according to neoclassical economic theory, if the private firm is operating as a monopoly, there will be productive inefficiencies and welfare loss through higher prices.

Based on the above arguments, we see that ownership of infrastructure assets may matter less for performance than the business environment of its operation and capital market discipline by lenders and investors (Vickers and Yarrow, 1991). It is by no means clear that the public sector is less efficient.

Infrastructure investment

Most countries invest around 2–8% of their annual Gross Domestic Product (GDP) on infrastructure, and the world average is around 4%. Based on the global nominal output of goods and services of US$100 trillion in 2022, this works out to about US$4 trillion a year, or US$40 trillion over the next decade. These investments may consist of greenfield (new) development or brownfield redevelopment of existing sites or facilities.

Currently, because of poor economic performance, large fiscal debt, war, and other reasons, many countries underinvest in infrastructure. The resulting shortfall is about US$350 billion a year over the next decade (McKinsey Global Institute, 2016). This is unfortunate because infrastructure investment confers many benefits such as

- lower transport and communication costs;
- greater production because of better access to markets and inter-firm linkages;
- reduction of spoilage of inputs, intermediate goods, and outputs;
- sharing of ideas through information technology and communications infrastructure;

- greater access to financing by helping to develop domestic capital markets;
- income and employment generation;
- improving access to facilities, education, healthcare, and job opportunities for the poor;
- reduction in regional inequalities;
- short-run stabilization of the economy through Keynesian demand management during recessions and, conversely, the reduction in public infrastructure spending during economic booms to curb inflation;
- more sustainable development by reducing traffic congestion, pollution, and promoting the use of renewable energy; and
- more resilient development by mitigating the adverse effects of climate change and viral attacks.

Some of the considerations are political, such as regional development, access for the poor, income and employment generation, and economic stabilization. Hence, it is possible to over-invest in infrastructure, such as building airports, bridges, and hospitals in declining regions that do not require them. Diminishing returns may also set in, resulting in lower rates of return. Diminishing returns mean that increasing an additional factor of production results in smaller increases in output. For example, all else being equal, if more and more tractors are added to a farm, the increase in output becomes smaller and smaller. There are just too many tractors, assuming that the farm size and number of workers are fixed, that is, all else being equal. Finally, with the increase in infrastructure assets over time, more funds will have to be spent on maintenance and repairs.

Infrastructure as an asset class

Infrastructure is an asset class because it generates stable and predictable long-term cash flows, which makes it attractive to public and institutional investors looking for opportunities to diversify their investment portfolios and hedge against inflation. For example, insurers will be able to match their long-term assets and liabilities by investing in infrastructure. Similarly, infrastructure investment is a critical part of the investment strategy of sovereign wealth funds such as the Government of Singapore Investment Corporation (GIC).

The hedge against inflation arises because output prices, such as road tolls and tariffs of utilities, tend to be indexed to inflation. Often, these tariffs have pass-through arrangements to pass rising input prices (e.g. oil) to consumers. If output prices are denominated in local currency, there is often provision to adjust for foreign exchange fluctuations.

The investment returns are largely unrelated to fluctuations in the short-term business cycle because of the long-lived nature of infrastructure assets. For example, the dividend yield of Keppel Investment Trust (KIT) has been relatively stable despite the COVID-19 pandemic (Table 1.1).

Table 1.1 Dividend yield of KIT.

Year	Dividend yield (%)
2020	5.21
2019	6.95
2018	6.95
2017	6.95
2016	5.21

Infrastructure such as roads and utilities tend to have monopoly or quasi-monopoly characteristics. Hence, there are barriers to entry. Infrastructure is also protected from demand obsolescence. Demand is generally price inelastic. This means that the quantity demanded is not sensitive to price changes because consumers still need to travel and use water, electricity, and gas. Owners and operators of such infrastructure have greater flexibility if they have price control. In other cases, the prices may be regulated by the appropriate public agency.

Finally, infrastructure assets are less prone to technological obsolescence than digital networks. However, nothing is static; even roads compete with air, sea, and rail travel. Newer forms of transport, such as air taxis, autonomous vehicles, and drone deliveries, will shape the future of transport.

According to the MSCI World Infrastructure Index, the annualized gross return is about 5.72% over the past decade (2012–22). Infrastructure projects tend to have different risk-return profiles (Table 1.2). At the lower end is social infrastructure, such as courthouses and schools, where the

Table 1.2   Risk-return profiles of infrastructure projects.

Return	Risk		
	Low	Medium	High
Low	Social infrastructure		
Medium	Road transport, utilities		Rail transport
High			Energy, airports, seaports

risks and returns are relatively lower. In contrast, energy, airport and seaport projects tend to provide higher returns but are riskier. For example, airports are highly competitive and prone to disruptions from bad weather, natural disasters, terrorism, pandemics, and labor strikes. Singapore, Kuala Lumpur, and Bangkok are competing to be the airport hub of Southeast Asia. All three cities have been busy expanding their airport capacities.

There are many ways to invest in infrastructure assets. It may be direct through project equity participation or indirect through infrastructure investment funds, real estate investment trusts (REITs), and securitized infrastructure loans. To securitize a loan or asset is to convert its value as sellable securities that are then sold to investors. For the owners of these loans or assets, securitization frees up tied-up capital. Unlike home mortgages, where securitization comes with higher risks of default because of changes in mortgage interest rates (Shiller, 2008), infrastructure assets, when pooled, tend to provide more stable returns. To encourage subscription, fund managers pool infrastructure assets from within a sector or from different sectors to increase diversification. These assets are in different stages of operation so that when the dividends are stapled (i.e. tied together), they provide fairly stable returns.

Long waves of infrastructure investment

Some analysts claim the existence of "long waves" of infrastructure investment that are less regular than post-World War II short-term business cycles and with a period of about 40 to 60 years (Van Duijin, 1983). Business cycles are much shorter, such as the OPEC oil hikes of 1973 and

1979, the global recession of 1985, the Asian financial crisis of 1997–98, the subprime mortgage or global financial crisis of 2007–08, and the COVID–19 pandemic of 2020–22.

These waves have many possible causes, such as

- the replacement of long-lived infrastructure (Kondratiev, 1935);
- clustering of technological innovation (Schumpeter, 1939);
- institutional changes in the structures of capital accumulation (Aglietta, 1979; Bowles *et al.*, 1983);
- long-term variations in the profit rate (Mandel, 1999);
- demographics (see (Tylecote, 1991)); and
- war (Goldstein, 1988).

The technologies that drive these waves include water power, textiles, iron, railways, steel, electricity, chemicals, and so on (Table 1.3). According to long wave theorists, the COVID-19 pandemic of 2020–22 marks the end of the 5th Kondratieff wave.

Table 1.3   Kondratieff waves.

Wave	From	To	Years	Technologies
1	1780	1845	60	Water power, textiles, iron
2	1845	1900	55	Railways, steel, steam
3	1900	1950	50	Electricity, chemicals, internal combustion engine
4	1950	1990	40	Petrochemicals, electronics, aviation
5	1990	2022	32	Biotechnology, digital networks, software
6	2023	?		Artificial intelligence, robotics, sustainability

There is no consensus on the theory and evidence on long waves. It is difficult to build a theory to explain long-term capitalist development in a country or on a global scale. Empirical evidence is also scant because of the need to observe long periods of upswings and downswings (Berry, 1991). Hence, the dates in Table 1.3 are debatable. Given the current geopolitical tensions between the United States and China, the Russia–Ukraine war, the climate change conundrum, the energy crisis, unprecedented demographic shifts, and the global debt crisis, it is difficult to believe that we are the start of a new Kondratieff upswing.

Sustainable infrastructure

Recall that infrastructure investment should be financially sustainable. This involves avoiding excessive investment and high operating and maintenance costs. Infrastructure projects should also be ecologically sustainable. This requires adherence to the four principles of sustainability, namely,

- the precautionary principle to safeguard the environment, particularly irreversible damage, if there are insufficient scientific evidence on the ecological impacts of projects;
- inter-generational equity so that future generations do not inherit a worse-off environment;
- conservation of biological diversity and ecological integrity so that different species can survive; and
- inclusion of environmental factors in valuing assets, such as the use of cost-benefit analysis (CBA) in evaluating projects.

The development of sustainable infrastructure requires the use of clean technology, that is, less energy, material, waste, pollution, and environmental damage.

On energy, governments may reduce carbon emissions by regulating, taxing, and trading them.

Carbon regulation involves "command and control" fixing, measuring, monitoring, and enforcing emission standards. This regulatory oversight using physical controls is complex, and requires considerable effort and trial and error. Further, firms have no incentive to cut back on allowable levels of pollution. Hence, economists prefer carbon taxes or carbon trading, although these measures have downsides as well.

A carbon tax will encourage major carbon producers to reduce their carbon emissions. Such taxes are often politically difficult to impose. Consequently, in most countries, the tax rate is modest and insufficient to spur substantial reductions in emissions. Producers may also threaten to pass the higher cost to "angry" consumers. For example, the Australian government imposed a carbon tax of A\$23 per ton in 2012 but reversed it the next year. Nonetheless, many governments have imposed the carbon tax. They use the tax revenue to fund projects to mitigate climate change or distribute it to consumers as carbon dividends to compensate for the higher energy costs.

In carbon trading, the government sets the annual cap on carbon emissions and allocates or sells these allowances. Firms that can cut back substantially on their emissions will have excess allowances, and they can sell them to firms that require these allowances. The annual cap is flexible, and carbon markets may be regional, national, or international. Like carbon taxes, "cap and trade" systems have their share of political difficulties. The main concern is the reluctance of many governments to impose costs on industry and consumers, resulting in weak caps. In turn, the large supply of allowances leads to low carbon prices. For example, in the European Union (EU) market, prices fell from €30 to €4 per ton between 2008 and 2013. Since then, the impact of climate change has forced a rethink on carbon trading. With stronger caps, EU carbon prices have climbed back steadily to around €25 in early 2021 before spiking to €100 by mid-2023.

The United Nations (UN) project-based Clean Development Mechanism (CDM) started in 2005 and works slightly differently from the "cap and trade" system. It allows governments and firms in rich countries to buy certified carbon credits (CCCs) from poor countries. This arrangement benefits the poor countries by encouraging the development of green projects to earn CCCs. The project originator sells these credits through a broker to finance part of the project. However, like the "cap and trade" system, participation in CDM fluctuates with carbon prices.

At the global level, efforts to reduce carbon emissions, such as the Paris Agreement (2015), tend to suffer from collective action problems. A country that does not agree with the terms of the agreement is unlikely to contribute towards the Green Climate Fund to reduce greenhouse gas emissions.

Green infrastructure

Green infrastructure means designing with nature to achieve ecologically sustainable infrastructure (Dover, 2015). For example, many countries have green building programs that evaluate the sustainable features of renovated or new buildings. These rating systems award points for

- location and orientation;
- water efficiency and harvesting;
- energy and atmosphere;

- use of natural ventilation;
- materials, resources, and recycling;
- indoor environmental quality;
- maintainability of buildings; and
- resilience (Reeder, 2010).

Beyond individual buildings, green infrastructure may be contrasted with gray infrastructure, such as the traditional system of gutters, pipes, and stormwater drains that collects and sends rainwater to local water bodies. Green infrastructure considers

- the use of rain gardens, bioswales, wetlands, diversion canals, and permeable surfaces to retain or reduce such flows to prevent flooding, keep the city cool, and create pleasant green spaces;
- access to public transport and alternative modes of mobility;
- the restoration of natural habits;
- local sourcing of materials and labor;
- the use of rapid renewable products such as bamboo;
- the project's impacts on ecology and local communities; and
- life cycle costing (LCC).

LCC is the present value of the cost of a physical investment. An equipment may have a low initial cost but higher operating and other costs. These costs should be compared and discounted to the present value to ascertain the LCC. For example, consider a machine that lasts four years with the following characteristics as shown in Table 1.4.

Table 1.4   Data for computing LCC ($).

Costs	0	1	2	3	4
Initial	2,000				
Operating		400	400	400	400
Maintenance		100	100	100	100
Repair				300	
Disposal					0
Salvage					0
Total	2,000	500	500	800	500

Then, using a 5% discount rate, the LCC is

$$L = 2,000 + \frac{500}{1.05} + \frac{500}{1.05^2} + \frac{800}{1.05^3} + \frac{500}{1.05^4} = 4,052.$$

A machine with a lower value of LCC is preferred.

A *green bond* is a fixed-income financial instrument that is used to fund projects that have positive environmental or climate benefits. A *climate bond* is issued to deal with carbon emissions. Similarly, a *blue bond* raises funds to improve fragile ocean ecosystems, reduce pollution, and support fisheries. Governments tend to offer tax advantages on these bonds to encourage investment in green infrastructure. However, bond issuers are *greenwashing* if they make exaggerated environmental claims.

Resilient infrastructure

A resilient infrastructure system is able to function during and after a disaster or adapt to climate change. Such a system will avoid expensive repairs after a hazard, reduce disruptions to economic and social activities, improve human well-being, and minimize human injuries and fatalities.

There are two ways to promote the development of resilient systems. The management approach (Gardoni, 2020) includes

- modeling the impact of climate change and hazards on infrastructure;
- better maintenance and repairs;
- developing effective emergence responses;
- developing a disaster mitigation plan; and
- modeling the recovery of physical and socio-economic systems.

The design or planning approach (Infield *et al.*, 2019) focuses on

- altering urban form, such as to reduce transport costs or the urban heat island effect;
- technical design of physical systems, including building redundancies;
- proper siting of infrastructure systems;
- use of protective landscaping;
- building resilience in project procurement;

- promotion of more self-sufficient regions to mitigate serious supply disruptions; and
- recognizing that infrastructure systems may be inter-dependent.

Both approaches require effective institutions to support the design, management, and implementation of resilient measures. Since the impact will be uneven across different regions and social groups, an inclusive and just strategy is also required to build resilient infrastructure (Sarte and Stipisic, 2016).

Inter-dependency means that a disruption in one system affects another. For example, during hot and dry summers, there are often droughts, and the hydropower dams may not produce sufficient electricity. This comes at the time when electricity demand peaks as households turn on their air-conditioning.

Stakeholders

There are many stakeholders in a public–private partnership (PPP) infrastructure project, such as

- politicians who support or oppose the project;
- the Ministry of Finance, for project approval and funding the public sector share such as periodic payments for the use of a facility;
- the relevant ministry, agency, or *grantor*, such as the Land Transport Authority, that awards the PPP contract to the winning bidder;
- other public agencies such as the Urban Redevelopment Authority, Land Authority, and National Environmental Agency that regulate, approve, support, or coordinate project proposals and processes;
- *sponsors* who provide equity and set up the project company or special purpose vehicle (SPV) to develop the project after signing the PPP contract with the grantor;
- lenders, including commercial banks, multilateral development banks and import-export banks;
- infrastructure funds, credit rating agencies, and other financial institutions;
- insurers;

- affected businesses, landowners, and residents;
- project, legal, design, engineering, financial, and other consultants;
- contractors and subcontractors;
- suppliers;
- workers and their unions;
- users, such as rail commuters or purchasers of water, electricity, and gas; and
- non-government organizations (NGOs).

For cross-border projects such as high-speed rail, foreign governments are likely to be involved.

Complexity

The provision of infrastructure is embedded in a complex system of stakeholders. It is not a linear democratic model where the elected representatives who form the government decide and the bureaucracy implements. The complexity lies in dealing with multiple interacting stakeholders who have different definitions of the problem, ideas, interests, power, and resources (Morcol, 2012). In other words, there is diversity, high interconnectedness, and instability in agent behavior. These three properties are the hallmarks of a complex system.

There are two main ways to deal with complexity. The first approach uses tight planning and control, or traditional project management. The second way is to learn as you go, such as the use of agile project management (see Chapter 2).

Environmental, social, and governance framework

The environmental, social, and governance (ESG) framework has been around for decades in different forms. For example, traditional cost-benefit analysis considers these three dimensions. The environmental aspects have been discussed under green infrastructure. The social dimension includes fair wages, equal opportunity, and positive impacts on the local community.

Governance refers to the decision-making authority in a project; that is, who gets to participate and decide. For the World Bank (1992; 2002), *good governance* includes

- public accountability and transparency;
- regulatory quality;
- the rule of law;
- control of corruption;
- democratization, decentralization, and local government reform;
- civil society and inclusive participation; and
- respect for human rights and the environment.

Unlike the free market approach of the 1980s, the Bank recognizes that development is more complex than just "getting prices right" or implementing "good polices." Markets and governments are now seen as complementary. Governments need to design and implement good institutions to allow markets to function better by correcting for imperfect or asymmetric information, improving competition, reducing externalities, enforcing property rights and contracts, and correcting for market distortions that reduce efficiency and investment. In particular, the financial system requires prudent regulation because banks may take on excessive risks, and the system is prone to bank runs and financial collapse on a global scale.

The Bank advocates administrative and fiscal decentralization to local governments that are more responsive to local demands for public goods and services. It brings the state closer to the people. The political decentralization of authority and responsibilities (called devolution) is more difficult as it involves the issue of power. However, local governments may be captured by local elites and are hence not democratic or responsive (Cheema and Rondinelli, 2007).

The outputs of infrastructure projects are also complex. They comprise many interacting civil, chemical, mechanical, and electrical components as well as processes. The consultants will need to coordinate the design, interfacing, installation, and commissioning of many sub-systems.

Challenges

Providing infrastructure is often a challenge for governments. The common problems often relate to

- governance;
- strategies and objectives;
- project governance structure;
- assessment of social, economic, and environmental performance;
- financing and funding;
- land acquisition;
- project selection;
- risk allocation;
- planning and coordination;
- construction; and
- operation and maintenance.

As discussed in the previous section, governance issues arise because stakeholders have different goals.

The project governance structure will have to be decisive, inclusive, and participatory. It will also involve the private sector under a PPP arrangement. There has to be "buy in" from social groups, such as the poor, the minorities, and the vulnerable, who will be impacted by the project. These groups should be empowered and are expected to coordinate their actions with NGOs. These NGOs will play a larger role in countries where state capacity is weak. Some bureaucrats think that public consultation is time-consuming and takes their precious hours away from projects with tight schedules. Further, there may be the perception that social groups tend to ask for the sky, well beyond the public budget. Despite these misgivings, public consultation is an established principle of good governance.

The public sector does not have all the skills and information to assess the social, economic, and environmental impacts of a project. It will need to hire external consultants to fill in the gaps.

Financing refers to the structuring of financial flows to a project, such as in arranging a loan. Funding refers to the transfer of financial resources, such as from the central to local government. Generally, PPP projects are financed by the private sector. The public sector's share lies in purchasing the project output, such as the use of the facilities. Since many local governments cannot raise sufficient revenues from property and local sales taxes, they require higher government grants.

The land acquisition process needs to be fair and efficient to avoid contentious challenges by affected parties. There may be long delays in settling the disputes because of unclear property rights.

Project selection should be based on national or local priorities. The evaluation of benefits and costs should be based on evidence, and there should not be a bias for new-build over rehabilitation projects.

The government has to allocate the risks efficiently and equitably and not saddle the private sector with excessive risks. The basic principle is to allocate a risk to the party that is best able to manage it. Generally, the government handles non-commercial risks, such as changes in law and regulation, while the private sector manages the commercial risks, such as demand, construction, and financial risks.

The public sector needs to strengthen its capacity to plan and coordinate projects through staff training and adequate provision of resources. Similarly, public agencies need to integrate their infrastructure investment plans into the national and city master plans and coordinate the implementation (Tan, 2023).

There are different *project delivery models*, such as

- direct provision by central government units;
- provision by lower government levels or agencies;
- private provision; and
- PPPs.

These choices affect the project funding, coordination, risks, design, construction, and operation. In this book, project delivery models are distinct from *procurement strategies* involving the selection of contractors, the type of construction contracts, and payment methods. The former concerns how the *public* sector intends to provide a good or service; the latter refers to how the *private* sector procures goods and services in a project. Although there is no consensus on the use of these terms, it is useful to make this distinction when dealing with PPP projects because there are two levels of delivery or procurement decisions.

Finally, the operation and maintenance (O & M) of infrastructure may be direct, delegated, privately executed, or carried out under a PPP arrangement. The traditional division of labor between central line ministries and lower levels of governments can cause O & M problems if the latter are poorly funded and not consulted for design or construction inputs.

References

Aglietta, M. (1979) *A theory of capitalist regulation*. London: Verso.

Banfield, E. (1958) *The moral basis of society*. New York: Free Press.

Berry, B. (1991) *Long-wave rhythms in economic development and political behavior*. Baltimore: Johns Hopkins University Press.

Bowles, S., Gordon, D., and Weisskopf, T. (1983) *Beyond the wasteland*. New York: Anchor Books.

Cheema, S. and Rondinelli, D. (Eds.) (2007) *Decentralizing governance*. Washington DC: Brookings Institution Press.

Coase, R. (1937) The nature of the firm. *Economica*, **4**(16), 386–405.

Dover, J. (2015) *Green infrastructure*. London: Earthscan.

Frydman, R., Gray, C., Hessel, M., and Rapaczynski, A. (1999) When does privatization work? *Quarterly Journal of Economics*, **114**(4), 1153–91.

Fukuyama, F. (1995) *Trust*. New York: Free Press.

Gardoni, P. (Ed.) (2020) *Routledge handbook of sustainable and resilient infrastructure*. London: Routledge.

Ghate, R., Jodha, N., and Mukhopadhyay, P. (Eds.) (2008) *Promise, trust, and evolution: Managing the commons of South Asia*. New York: Oxford University Press.

Goldstein, J. (1988) *Long cycles: Prosperity and war in the modern Age*. New Haven: Yale University Press.

Hyden, G. (2012) *The economy of affection*. London: Cambridge University Press.

Infield, E., Abunnasr, Y., and Ryan, R. (Eds.) (2019) *Planning for climate change*. London: Taylor and Francis.

Kondratiev, N. (1935) The long waves in economic life. *Review of Economic Statistics*, **17**(6), 105–15.

Mandel, E. (1999) *Late capitalism*. London: Verso.

McKinsey Global Institute (2016) *Bridging global infrastructure gaps*. San Francisco: MGI.

Morcol, G. (2012) *A complexity theory for public policy*. London: Routledge.

North, D. (1990) *Institutions, institutional change and economic performance*. Cambridge: Cambridge University Press.

Orru, M., Biggart, N., and Hamilton, G. (1996) *The economic organization of East Asian capitalism*. New York: SAGE.

Ostrom, E. (1990) *Governing the commons*. New York: Cambridge University Press.

Putnam, R. (2001) *Bowling alone*. New York: Simon and Schuster.

Reeder, L. (2010) *Guide to green building rating systems*. New York: Wiley.

Sarte, S. and Stipisic, M. (2016) *Water infrastructure: Equitable deployment of resilient systems*. New York: Columbia University Office of Publications.

Schumpeter, J. (1939) *Business cycles*. New York: McGraw-Hill.

Shiller, R. (2008) *The subprime solution*. New Jersey: Princeton University Press.

Tan, W. (2023) *Urban management: Managing cities in uncertain times*. Singapore: World Scientific Publishing.

Tylecote, A. (1991) *The long wave in the world economy*. London: Routledge

Van Dujin, J. (1983) *The long wave in economic life*. London: George Allen & Unwin.

Vickers, J. and Yarrow, G. (1991) Economic perspectives on privatization. *Journal of Economic Perspectives*, **5**(2), 111–32.

World Bank (1992) *Governance and development*. Washington DC: World Bank.

World Bank (2002) *Building institutions for markets*. Washington DC: Oxford University Press.

CHAPTER 2

The stages of infrastructure development

Institutional framework

Projects are embedded within often taken-for-granted institutional frameworks so that they can be properly managed. Institutions are the rules of the game or the players themselves, such as organizations.

The elements of this institutional framework include government, political systems, organizations, markets, regulatory systems, financial systems, property rights, the legal system, and the social system (World Bank, 2002). History and geography play an important role in the development of institutions; for example, ex-colonies inherit different sets of institutions that set them on different growth trajectories (Acemoglu and Robinson, 2012). Countries with weak institutions need to build effective ones to support projects and generate economic growth.

It is assumed that the public agency or organization has the capacity to finance, design, and implement projects. The capacity of a government depends not only on its internal resources, processes, and institutions, but also on the relation between the state and society that defines its capacity. In other words, it depends on the constellation of political forces in societies divided by class, gender, religion, race, regions, and other divisions (Tan, 2023). For cross-border projects, geopolitics will also affect project outcomes.

The project cycle

Before discussing the infrastructure development cycle, it is useful to consider the shorter construction project cycle, which consists of the following phases (Bennett, 2003):

- pre-project;
- planning and design;

- contractor selection;
- mobilization;
- site execution; and
- project close-out.

A slight variation is the following (Baum, 1982):

- project initiation;
- planning;
- execution;
- close-out; and
- operation and maintenance.

The main difference is that operation and maintenance is considered an additional phase after project close-out.

For other sectors, the project cycle is similar. For example, the software development cycle consists of (Philips, 2010)

- requirements gathering and analysis;
- planning and design;
- development;
- testing; and
- operation and maintenance.

There are variations, such as having a deployment phase after software testing or by combining both activities in a single phase.

Phases in infrastructure development

The infrastructure development cycle is much longer than the standard project cycle. A public–private partnership (PPP) project consists of the following phases:

- project identification;
- appraisal of options;
- feasibility study;
- project preparation;
- tender;

- sponsor's bid preparation;
- bid evaluation and contract award;
- lender's due diligence;
- sponsor's pre-construction activities;
- mobilization;
- construction or execution;
- project close-out;
- operation and maintenance; and
- handing over.

The grantor or contracting public agency, such as the Public Utilities Board, executes the first five phases, from project identification to the issue of tender. The first two phases comprise the pre-feasibility study.

The private sector then prepares and bids for the project. During this phase, the sponsors will need to seek financing. Lenders, in turn, will carry out their *due diligence* before granting in-principle approval of the loan to enable the bidder to tender for the project. Once the PPP contract has been awarded, the winning bidder will set up a project company or special purpose vehicle (SPV) to mobilize resources and construct the facility. Upon the completion of construction, the project enters the operation and maintenance (O & M) phase. Finally, at the end of the concession or contract period (e.g. 25 years), the facility is transferred to the public agency. This, in brief, is the common build-operate-transfer (BOT) PPP contract arrangement. There are many variations, as discussed in Chapter 3. The book follows this procurement strategy.

We will discuss each phase in greater detail in subsequent chapters. Each phase consists of many sub-stages. For example, the tender phase consists of pre-qualifying potential bidders, asking the shortlisted parties to respond to a Request for Proposal (RFP), responding to queries by bidders, and the issuance of the final tender after taking into consideration feedback from potential bidders.

Like the standard project cycle, there are many variations of the infrastructure development cycle, and projects may not have all the above phases. Some projects terminate at the end of the construction phase. For example, private developers may build housing units for sale and are not involved in the O & M phase. Similarly, internal projects within an organization may not involve bidding, financing, and transfer of assets at the end of the O & M period.

Traditionally, public agencies do not use the PPP delivery strategy. The private sector is only involved in building or managing the facility. The public sector finances the project. For example, many schools are still built using this approach. The private sector builds the schools through public tender. The education ministry then appoints a private operator to manage the facilities.

It is possible to terminate a project before the end of the PPP contract period. It can occur because of policy changes, default of contracting parties, prolonged *force majeure* events, achievement of a target profit margin, unlocking of asset value, securitization of assets, and so on.

Approaches to project management

There are two basic approaches to managing projects (Ajam, 2018). The traditional approach, common in infrastructure projects, is based on the *blueprint*, linear, or waterfall project cycle discussed in the previous section:

Requirements	Design	Construct	Test	Deliver

Each stage is completed before the next stage begins. For example, the gathering of requirements and design is largely completed before construction begins. The main reason for using this approach is that, once built, the physical infrastructure is difficult or costly to change. Subsequent changes are made using variation orders. One can also take a *system view* of project management and conceptualize sub-systems of project initiation, planning, design, control, and project close-out (Roman, 1986).

In the information technology (IT) and manufacturing industries, the *agile* approach is more common because it is easier to make changes (Brechner, 2015). There are different versions of agile project management. The basic idea is to prioritize the user or owner requirements and then rapidly develop ("sprint") a prototype for delivery to the owner for further discussion (Fig. 2.1). During the sprint, which takes less than a month, the project team holds daily meetings or "scrums" to iron out the

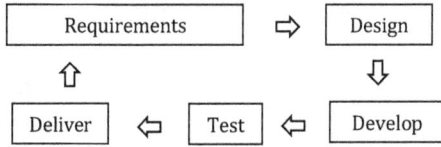

Fig. 2.1   Agile project management.

issues. The process is continuous as more requirements (features) are added at each successive sprint until the product is fully developed.

Agile project teams may use the *Kanban* framework to monitor progress. It is basically a tracking table:

Done	To do	In progress	Stories

Stories can be in the form of comments, such as the causes of delays. Some project managers treat scrum and Kanban as different methodologies (Kniberg and Skarin, 2009). It is better to flexibly integrate them as discussed here.

Leadership and teamwork

Leadership is often seen as important in the business world. As a concept, it is vague despite many years of research, which is why it is difficult for organizations, no matter how hard they try, to spot or train leaders. The concept is further muddied by differentiating leaders who set the vision and managers who execute the plan (Bennis, 1989), and calling the leader of the organization the Chief Executive *Officer* (CEO), which is a military term. Historically, a general was not a leader; as an officer, he had "general command" over the king's army, that is, a commander. Finally, business schools do not train leaders; their main product, the MBA, stands for Master in Business *Administration*, not leadership.

Early debates on leadership centered on whether leaders are born or made (de Vries, 2004), which gets nowhere. If leaders are born leaders, there is no point in training leaders. If leaders are made, it begs the question of how to transform a person into a leader. There are many claims on how one *becomes* a leader (Munroe, 2018), but none is definitive.

Subsequent thinking focuses on the intangible qualities a leader *has*, that is, traits (Davis, 2010). It is also known as Great Man Theory (Hook, 1955), with men as heroes in *his*tory, not women. History is the story of lions, not deer. However, there is a long list of traits a leader should possess, such as physical appearance, self-efficacy, wisdom, ability to see the big picture, focus, positivity, empathy, fairness, courage, intelligence, creativity, confidence, trust, morality, superior decision-making, risk-taking, adaptability, cultural sensitivity, and the ability to network well. Leaders are thought to have different styles, such as autocratic, democratic (participatory), or anything goes (*laissez-faire*).

A different way of looking at leadership is to consider what a leader *does*. This is the behavioral approach. Here, the leader has vision, persuades others of the benefits of following the leader, builds the system, strategizes, gathers resources, builds trust, motivates, monitors progress, and handles conflicts. In brief, the leader is the system builder (Hugos, 2004) as well as the team builder (Erwin and Devoll, 2021) or healer (Janni, 2022). However, most of the functions listed here are performed by managers, not leaders. Further, if leaders do different things, there is no core set of things a leader should do.

Perhaps what a leader does depends on the situation or context, especially the power structure in the organization. This approach is called contingent leadership. Other than power relationships, the task structure matters (Fiedler, 1967). If the tasks are highly structured, it is easier for the leader to monitor performance.

A final line of leadership research looks at what business leaders *say* about how they did it *my* way. There are many popular business books about how the CEOs of large corporations such as IBM (Gerstner, 2003), Walt Disney (Iger, 2019), or General Electric (Welch, 2005) turned the ship around, to their credit. Unfortunately, bits and pieces of CEO advice do not constitute a theory of leadership. Even dictators have a thing or two about doing things *my* way (Sikder, 2019).

Projects are carried out by different teams. The grantor or public agency has its project team to carry out the first five phases of the infrastructure development cycle and subsequently monitor the performance of the winning bidder. The project company has a project team to carry out the bid preparation and subsequent phases. Lenders will also have a project finance team to conduct due diligence and administer the loan. Finally, the contractor and subcontractors also work in teams, and these teams may be split into office and site groups.

Unlike large corporations, these are small teams, and it does not make much sense to think of the project manager as the leader of the team. Nearly every project relies on team effort.

Each team should have a balance of expertise and experience. These are the hard skills. Getting the team to work together requires soft skills. These "people skills" include personal skills such as the ability to lead, communicate, self-motivate, and adapt to changes (Bolton, 1986). The team should be able to set goals, strategize, plan, communicate, mobilize resources, motivate or incentivize others, solve problems, manage conflicts, and build consensus. Communication skills refer to the ability to listen, read, and write, as well as understanding non-verbal forms of communication. A good work ethic and the ability to self-motivate are also important traits. Finally, one has to be flexible and adapt to changes. There will be many changes in a project, and we should not be fixated with our ideas or attitudes.

People skills also include inter-personal skills, or how team members relate to one another. The ability to collaborate with others will depend on having a positive attitude, being friendly and able to empathize, being diplomatic in giving feedback, and being able to take criticism well.

Academics tend to distinguish the cognitive, psychomotor, and affective domains. The cognitive domain involves the mind or understanding, which is knowledge. The psychomotor domain involves the body and, hence, skills. For example, one may be knowledgeable about football but not have the skills of a professional footballer. Finally, the affective domain involves matters of the heart, or our attitudes. In brief, this classification involves the mind, body, and heart. The industry reduces it to hard and soft skills. These classifications may cause confusion over the definition of skills, such as Ralph Marston's remark that "excellence is not a skill, it's an attitude."

Integration

Traditionally, infrastructure development consists of separate economic, financial, and other analyses carried out by different agencies.

Nowadays, the process is integrated during the different phases of development (Sell, 1991). For example, an integrated feasibility study includes political, technical, economic, financial, fiscal, environmental, and social considerations. A second aspect of integration consists of internal and external stakeholders working together to achieve development objectives.

Unlike infrastructure development, the project management literature conceptualizes project integration differently. One approach focuses on *project integration management* (PIM) as the coordination of seven different processes (Cioffi, 2002): procurement, scope, time, cost, quality, communication, and risk. This is achieved by developing a project brief, developing baseline plans, assigning roles and responsibilities, managing and controlling the execution, performing integrated change control, and closing out the project.

Another approach is integrated project management (IPM) (Barkley, 2006), which consists of integrating

- customer requirements;
- design and construction, through a multi-party contract among team members;
- technologies, through interfaces;
- the execution of multiple projects, through program management;
- system support, through integrative organizational designs; and
- financials and schedule, through earned value analysis.

The lesson is that all these processes require integration in complex infrastructure projects.

References

Acemoglu, D. and Robinson, J. (2012) *Why nations fail.* New York: Currency.
Ajam, M. (2018) *Project management beyond waterfall and agile.* London: CRC Press.

Barkley, B. (2006) *Integrated project management*. New York: McGraw-Hill.

Baum, W. (1982) *The project cycle*. Washington DC: World Bank.

Bennett, L. (2003) *The management of construction*. London: Butterworth.

Bennis, W. (1989) *On becoming a leader*. New York: Basic Books.

Bolton, R. (1986) *People skills*. Greenwich: Touchstone.

Brechner, E. (2023) *Agile project management with Kanban*. Seattle: Microsoft Press.

Cioffi, D. (2002) *Managing project integration*. Vienna: Management Concepts.

Davis, R. (2010) *The intangibles of leadership: The 10 qualities of superior executive performance*. New York: Wiley.

de Vries, M. (2004) *Are leaders born or are they made? The case of Alexander the Great*. London: Routledge.

Erwin, M. and Devoll, W. (2021) *Leadership is a relationship*. New York: Wiley.

Fiedler, F. (1967) *A theory of leadership effectiveness*. New York: McGraw-Hill.

Gerstner, L. (2003) *Who says elephants can't dance?* New York: Harper Business.

Hook, S. (1955) *The hero in history*. Boston: Beacon Press.

Hugos, M. (2004) The systems builder as leader. *Computerworld*, 9 August.

Iger, R. (2019) *The ride of a lifetime: Lessons learned from 15 years as CEO of the Walt Disney Company*. New York: Random House.

Janni, N. (2022) *Leader as healer*. London: LID Publishing.

Kniberg, H. and Skarin, M. (2009) *Kanban and scrum*. Morrisville: Lulu Press.

Munroe, M. (2018) *Becoming a leader*. London: Blackwell.

Philips, J. (2010) *IT project management*. New York: McGraw-Hill.

Roman, D. (1986) *Managing projects: A systems approach*. Amsterdam: Elsevier.

Sell, A. (1991) *Project evaluation: An integrated financial and economic analysis*. Aldershot: Avebury.

Sikder, A. (2019) *My way! Leadership styles of history's 18 most prominent dictators*. Haryana: Invincible Publishers.

Tan, W. (2023) *Urban management: Managing cities in uncertain times*. Singapore: World Scientific Publishing.

Welch, J. (2005) *Winning*. London: Thorsons.

World Bank (2002) *Building institutions for markets*. Washington DC: World Bank.

CHAPTER 3

Project identification

Sources of projects

Projects arise from perceived problems, social needs, or investment opportunities. Many projects originate from government infrastructure investment plans found in national, regional, sectoral, and other strategies. In this sense, projects are the manifestations of economic and business strategies.

There are also international projects, such as those developed under China's Belt and Road Initiative (BRI), to create a huge international market for goods, services, talent, ideas, and capital (Griffiths, 2017). Announced in 2013, it involves infrastructure investments in Asia, Middle East, Europe, Africa, and the Americas over many decades. The "Belt" refers to overland road and rail transport, and the "Road" refers to the maritime sea routes. Other types of cross-border projects include oil and gas pipelines, high-speed rail, power projects, road tunnels, and submarine cables.

Some governments are open to unsolicited proposals from the private sector. These proposals may just be initial or innovative ideas and require careful screening to avoid perceptions of corruption or its sensitive or confidential nature. If a proposal is promising, there is the issue of whether to use direct negotiation, such as on a cost-plus-fee basis or competitive tender (World Bank, 2019). In the latter, the proposer may be automatically shortlisted as a bidder, given bonus points in tender evaluation, or given the option to match the winning bid and be awarded the PPP contract. Like all proposals, there is also the issue of fiscal support throughout the duration of the PPP contract.

Infrastructure projects can stand alone or form a part of a series of related projects called a *development program*. For example, a rural integrated development program may consist of the building of institutions, roads, schools, medical facilities, markets, irrigation systems, and so on (Hebbar, 1991). Similarly, a long-term housing development program may consist of the building of a series of new towns and housing estates. Unsolicited proposals may not integrate well in such programs.

Project goals and objectives

Regardless of the sources, projects have stated goals and objectives. Typically, from the government's viewpoint, broad goals include connecting one urban settlement to another, expanding agricultural output, alleviating poverty, improving public education, building resilience, improving public safety and health, and developing a better and more sustainable living environment. The objectives are more specific: expanding agricultural output by constructing an irrigation system, preventing soil erosion, and carrying out reforestation.

These objectives should relate to project outputs rather than inputs. For example, the purpose of a high-speed rail (HSR) project should be to connect urban centers and not increase employment. The latter is an input, and the money not spent on the HSR project can be used to create employment in alternate projects.

Preliminary screening

Regardless of the sources of potential projects, an initial project or study team in the primary ministry ranks and screens them to eliminate undesirable ones. Some proposals are simply wild ideas that are unlikely to be taken seriously to save valuable time and cost.

Assessment of options

For projects that show promise, the next step is to consider possible options for each proposal.

One option is to do nothing, but only after evaluating other options by considering the demand and supply of infrastructure. On the demand side, it may be possible to reduce the demand for the infrastructure through appropriate pricing or some form of rationing. On the supply side, there is the option to upgrade the existing facility or build a new one.

More broadly, we need to compare the different states of nature "with and without" the project. This should not be confused with a "before and after" assessment where the world is assumed to be unchanging or static. For instance, suppose a city is currently experiencing traffic congestion during peak hours, and there is a proposal to build a new highway to deal with the problem. Even if the highway is not built, the traffic conditions are likely to worsen.

The state of nature without the project is also called the *counterfactual* or base case. This terminology is used in the classical experimental design (Tan, 2022):

$$
\begin{array}{ccccc}
E & R & X & T & Y \\
C & R & x & & y
\end{array}
$$

Here, E is the experimental group, and C is the control group. Both groups are randomized (R) to ensure comparability, and the experimental group is given the treatment (T). The symbols X, x, Y, and y are pre-test and post-test scores. The control group acts as the counterfactual. A simple example of this design is to select two similar car sales teams from two comparable cities, and T represents an incentive scheme implemented in E. Here, X, x, Y, and y are car sales before and after the experiment, such as over a year. Without the control group, we may incorrectly attribute higher car sales to T if $Y > X$. The control group may have similar growth in car sales, such as if both cities are experiencing an economic boom in the country.

In a project scenario, T is the project, and the analyst has to imagine what will happen without the project. For example, if T is an irrigation project, will agricultural output have increased without the project over the same period?

For bold new ideas, there is the option of piloting some of these development projects as policy experiments to test and learn from the experience (Rondinelli, 1993). This adaptive approach is similar to the agile project management route discussed in the previous chapter.

Preliminary study

Given the goals and objectives, the preliminary or pre-feasibility study appraises the various options for each potential project. For a new facility, it consists of

- technical assessment;
- economic appraisal; and
- political, regulatory, and legal assessment.

We discuss these items below.

Technical assessment

The technical assessment covers

- project scope and basic specifications;
- site options;
- basic design and engineering options;
- preliminary schedule;
- project governance structure;
- choice of project delivery; and
- procurement strategy.

A narrower definition of technical assessment restricts it to design and engineering evaluation. The project scope and basic specifications may be developed top-down or bottom-up by soliciting inputs from other stakeholders. These specifications include the target users, functional requirements, main materials, culture and aesthetics, environmental sustainability, resilience, and the desired performance of the facility. At this early stage, the specifications are preliminary and may consist of just the sizes or massing of the functional elements. For example, for a school, the functional components may include an administrative and staff block, parking, canteen, school hall, dormitories, classrooms, library, laboratories, basketball court, tennis court, football field, running track, and an educational garden.

The study team then evaluates the site, design, and engineering options. If a site is not yet available, the public agency will have to look for suitable sites. In many cases, it is likely to own a suitable site. Otherwise it may have to acquire it, such as when an existing road is being considered for upgrading. At this stage, it is unnecessary to acquire a site as the assessment is preliminary, and the project may not be feasible. If it is a private sector project, the developer is likely to purchase an option from the landowner to keep the land from the market, usually for about three months. If the project is subsequently found to be feasible, the developer will exercise the option to purchase the land.

Once a site has been identified, the team proceeds to develop a *conceptual design* for the site. The design is used to check if the functional elements can be accommodated. It may be necessary to build at a higher density because of a smaller site. The layout of the elements is important; for example, the entrance should have a unique focal point to identify the school, the car park should be near the entrance to reduce traffic movement, the classrooms and laboratories are usually stacked in one or more blocks, and so on. The idea is to minimize noise and the movement of goods and people. The buildings should be oriented appropriately to exploit the good views as well as harness the sun and wind to generate power and minimize energy consumption.

At this early stage, little is known about the project, and the preliminary schedule is just a simple bar chart with broad timings for the project phases and possible milestones. These phases, activities, and timings are obtained from experience in executing similar projects.

The *project governance* structure spells out who has decision-making authority in the project and the reporting of progress. It is a high-level decision-making body comprising senior management and key stakeholders. At this stage, the structure is evolving and may not be complete.

The choice of *project delivery* includes public provision, private provision, or a mix of both. Normally, the choice is made at higher levels; the study team does not make this decision. If the government provides the output, it may do it centrally or delegate it to lower levels of government. The government is likely to build a public park because it cannot exclude non-paying users. Similarly, it may build a public school and subsidize

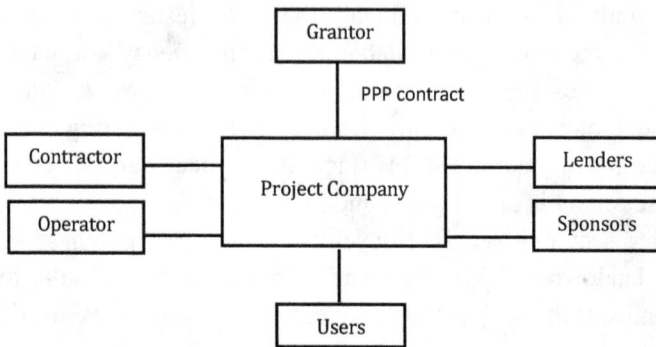

Fig. 3.1   PPP contract structure.

primary education alongside private schools. Finally, it may engage a private operator to manage the facilities.

In this book, the *procurement strategy* is based on a public–private partnership (PPP) contract (Fig. 3.1). The grantor (public agency) enters into a PPP contract with a special purpose vehicle (SPV) or project company. The SPV designs, builds, finances, operates, and maintains the infrastructure to specifications and earns revenues from it during the contract or concession period (e.g. 25 years). The SPV may upgrade an existing asset rather than build a new one.

Sponsors are active equity owners of the project company and conduct their businesses through a shareholders' agreement. They raise and contribute the initial equity and borrow the rest through a commercial loan agreement. The initial equity or risk capital may take the form of a sponsor loan to the SPV to reduce its tax liability. That is, interest on a loan is tax deductible.

Commercial lending is on *limited recourse*; that is, lenders have recourse to only project cash flows and assets if the project fails. This is sometimes called off-balance sheet financing or bankruptcy remoteness. Limited-recourse financing means that if the project fails, the corporate assets of the parent company are not at risk. For example, a developer will set up different project entities, one for each project, for limited-recourse financing. If a project fails, lenders have recourse to only the assets of the project entity. There is often limited support from sponsors or their parent companies, such as contingent equity and guarantees against cost overruns.

Lenders require sponsors to establish the SPV after the award of the PPP contract to "ring fence" the project assets, contracts, and cash flows as security for the loan.

The SPV then contracts with a contractor to build the facility and, upon completion, appoints an operator using an operation and maintenance (O & M) contract. It then sells the output through an off-take contract or directly to users. At the end of the contract period, the SPV may own the asset or transfer it to the government. The former is a Design-Build-Operate-Own (DBOO) arrangement, and the latter is a Build-Operate-Transfer (BOT) model. Both models are also called Design-Build-Finance-Operate (DBFO) arrangements with or without a transfer of assets depending on the residual value and whether the land is privately or publicly owned.

The two main PPP payment models are

- grantor pays; and
- user pays.

The "grantor pays" model has three variations, namely,

- availability basis;
- output basis; or
- input basis.

In general, the grantor pays for the "availability" of the facility in cases where there is little or no private demand, such as schools, prisons, sports facilities, and public hospitals. For example, the grantor pays for the use of the prison, irrespective of the number of prisoners.

The grantor may also be an off-taker that purchases the project output, such as electricity or desalinated water. Finally, the grantor may pay for the input, such as municipal solid waste in waste-to-energy (WTE) incinerator projects. Usually, the periodic amount to be paid to the SPV comprises fixed and variable components. The fixed costs (capacity charge) covers the development costs, debt service, taxes, and profit margin. The variable costs include energy, materials, operation, and maintenance.

Users pay for the project output or service if there is private demand, such as toll roads, bridges, and railways.

Governments have several reasons for using the PPP route. They may want to tap on private funds to overcome fiscal constraints in building infrastructure and achieve earlier project completion. PPP projects also provide better *value for money* (VfM) through

- competition and transparency in project bids;
- cost efficiency;
- private sector expertise and innovation;
- capacity building and opportunity to learn from the private sector;
- rigorous assessment of project feasibility;
- opportunities for risk sharing and transfer;
- whole life consideration of design, construction, operation, and maintenance;
- accountability and service performance by linking payment to performance;
- opportunities to share the use of facilities with other users to lower the total cost to the public;
- opportunities to generate and share third-party revenue; and
- possible transfer of asset to the grantor at the end of the concession period.

Some governments use PPP as an opportunity or catalyst for structural reforms such as privatization, deregulation, and development of the local financial sector in infrastructure lending, insurance, and financial management of risks. Global lenders such as the World Bank and International Monetary Fund (IMF) may require a government to implement structural reforms as a condition for loans.

The possible downsides to the use of PPP include

- higher financing cost of about 2 to 3%, assuming the government can borrow at a lower cost of capital;
- higher transaction cost of searching for information, negotiation, monitoring, and settling of disputes because of contractual complexity;
- possible user affordability issues if the SPV subsequently raises the price of output;

- safety, environmental, and service quality concerns because of the profit motive; and
- possible lack of participation by the private sector if the risks are perceived to be too high or improperly allocated.

The issue of whether the profit motive results in safety, environmental, and service quality concerns is debatable.

PPP projects tend to be more successful if there is a constant revenue stream, such as in utilities, or if the government is able to pay the annual fee on a sustainable basis, such as in building courthouses. It is less suitable for building costly stadiums and where the revenue stream is less certain for sporting events.

Economic appraisal

The second component of a preliminary study is economic appraisal, also known as cost-benefit analysis (CBA). It may be used to assess

- policy impacts, such as a tax policy (Levy, 1995);
- social impacts, especially on who gains and who loses (Derman and Whiteford, 2019); and
- project feasibility.

The focus of this section is project feasibility. The steps in CBA are:

- identification and valuation of benefits;
- identification and valuation of costs;
- decision criteria;
- risk assessment; and
- recommendation.

We discuss these steps below, followed by criticisms of CBA. The alternative to CBA is the project choice of the ruler or central planners. This choice may not benefit the public and may not be fair or efficient in allocating resources. Central planning has failed because planners lack knowledge of public preferences, incentives are perverse, and it is difficult to compute production quantities without prices (Hayek, 1976).

The outcomes are well-known: producing goods and services people do not want, nobody wanting to throw out the garbage, and frequent production excesses and scarcities.

Project benefits

A public project is economically feasible if it improves social welfare (SW). In turn, SW is assumed to be the sum of individual utilities. Only individuals count in the affected community, which means we cannot count the utilities of future generations, animals, and plants. However, we may make decisions or provisions for them, such as not to damage the environment.

Unfortunately, it is not possible to measure individual utility. A person may benefit from a quieter road, but we cannot measure how much utility is derived from it. Hence, CBA analysts use monetary values.

Benefit from marketable output

The benefit or value of an apple is based on a person's *willingness to pay* (WTP) for it. If a person is willing to pay \$2 for an apple, it measures the benefit from consuming it. Another individual may be willing to pay only \$1 for the apple. Hence, a project's benefit is revenue from its output, such as electricity or desalinated water. As shown in Fig. 3.2, the revenue is the area of the trapezium F. The project output is $Q^* - Q = q$. This increased

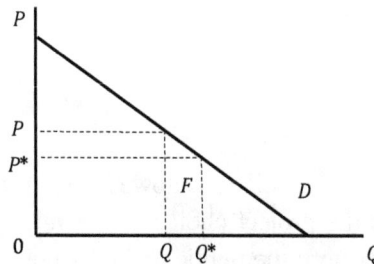

Fig. 3.2   Output benefit for a power generation project.

output may displace some less efficient producers because the price has fallen from P to P^*.

If the price fall is negligible, such as if a substantial part of the project output is exported and the firm is a price taker in the global market, then the demand curve D is relatively flat and F may be approximated by a rectangle.

If the price fall is large, the price elasticity of demand is given by

$$e = \frac{\% \ change \ in \ quantity \ demanded}{\% \ change \ in \ price}.$$

Rearranging,

$$\% \ change \ in \ price = \frac{q/Q}{e} \times 100.$$

If e, P, Q, and q are known, it is possible to find P^* and then use it to estimate F. It is sometimes assumed that $e = 1$; that is, an $x\%$ fall in price leads to an $x\%$ rise in quantity demand. This is a good approximation for many goods; for example, a 5% rise in prices tends to lead to a 5% fall in sales. The value of e may also be estimated from previous econometric studies, such as for electricity consumption (Shu and Hyndman, 2011). These studies are based on the equation

$$e = \frac{\% \ change \ in \ quantity \ demanded}{\% \ change \ in \ price} = \frac{\partial \log (Q)}{\partial \log (P)}.$$

The Delta (∂) symbol refers to the partial derivative. The partial derivative of y with respect to x is the derivative of y with respect to x, holding other variables constant. For example, if $y = x^2 + z + u$, then

$$\frac{\partial y}{\partial x} = 2x.$$

To understand the elasticity equation, recall that for any variable x,

$$d \log(x) = \frac{dx}{x}.$$

where dx is the differential. Hence,

$$\frac{d\log(Q)}{d\log(P)} = \frac{dQ/Q}{dP/P} = e.$$

The estimated demand equation is often written as

$$\log(Q) = a + b\log(P) + c\log(Y) + u.$$

Here, Y is the household income, and u is the residual term. Other demand shifters (variables) may be added to the equation, such as population growth, interest rates, taxes, expected rate of inflation, and the price of substitutes. Taking the partial derivative of $\log(Q)$ with respect to $\log(P)$ gives $e = b$. In other words, the coefficient b is the price elasticity of demand. This is a major reason why the demand equation is often estimated using the log function.

Benefit from non-marketable output

Not all project outputs are marketable, such as the building of schools or reforestation projects. In such cases, there is no simple measure of the benefit, and we minimize cost, such as the least-cost method of reforestation. If the outputs differ only in a single dimension, it is called *cost-effectiveness analysis* (CEA), such as the cost per pupil or cost per hectare.

Benefit from transport cost savings

In transport projects, the output or direct benefit is measured in terms of cost savings. Consider the case where a road is to be upgraded. For existing commuters making Q (round) trips a year, the cost of travel (including road toll) falls from C to C^* (Fig. 3.3). The benefit is the area X through savings in

- travel time;
- vehicle operating cost;

- injury cost; and
- property damage.

A second benefit is area Y, which represents the gain to new commuters who now travel because of the lower cost. The total benefit of the project is $X + Y$.

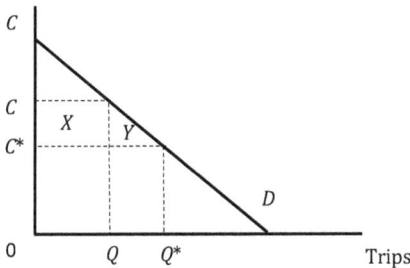

Fig. 3.3   Benefit for the transport project.

There are other short-term benefits, such as reduced congestion in other parts of the city because of the diversion of drivers to the upgraded road. There may be other costs as well; for example, the upgraded road may transfer congestion to another part of the city. We assume these effects are small, or they cancel out. There may also be longer-term indirect benefits, such as those that affect the quality of life. We ignore these benefits because they are often difficult to estimate and uncertain. Further, when discounted to present value at a relatively high discount rate, distant benefits are much smaller than short-term ones.

As before, the estimation of X and Y requires knowledge of the demand curve. In the short run, commuters driving to work are relatively insensitive to price changes (Goodwin *et al.*, 2004). However, unlike Fig. 3.2, it is more difficult to estimate Q^* in Fig. 3.3, which is why road toll projects are risky. The standard way of forecasting traffic demand is to consider

- trip generation;
- trip distributions;

- modal choice; and
- route assignment.

For example, if there are three zones (*A*, *B*, and *C*) in the city, the annual number of trips (in millions) are given below:

	Zone			
	A	*B*	*C*	Total
A	1	3	5	9
B	2	4	6	12
C	3	5	7	15
Total	6	12	18	36

The table is read from left to right. For example, the number of trips from *A* to *B* is 3 million, and from *B* to *A* is 2 million, as highlighted. The next step is to determine the model split in each cell. For example, of the 3 million trips from *A* to *B*, the modal split is as follows:

Mode	Trips
Bike	0.5
Car	0.5
Bus	1.5
Taxi	0.2
Rail	0.3
Total	3.0

The last step is to assign routes to each modal choice. For example, of the 0.5 million car trips, 0.3 million drivers will use the upgraded road. The main problem lies in getting the data, such as from household surveys.

A second method of forecasting traffic demand for the proposed upgrading is to extrapolate from the trend. For example, if traffic on the existing road increases by an average 3% a year over the last decade, we may use this information to forecast future trips.

A third method is to regress the annual number of trips on demand variables (shifters) such as per capita income and population growth. However, this method is not suitable for estimating demand for a particular road. It is better applied to estimating traffic growth for an entire city, region, or country.

Finally, it is possible to estimate a demand function that includes the price (cost) of travel and other demand shifters using data from other roads. As discussed earlier in this section, it is possible to estimate the price elasticity of demand by using a double-log functional form. However, in view of the quality and amount of data required, many analysts simply assume unitary price elasticity and use this information to estimate Q^* in Fig. 3.3.

Other benefits

Apart from output and cost savings benefits, a project may have other secondary benefits, such as an increase in the output of another industry. For example, a project to reduce soil erosion may stimulate additional farm production. The benefit (B) is the additional farm output (Q_i) for the ith crop multiplied by individual prices (P_i):

$$B = \sum_i P_i Q_i.$$

This is the *productivity change method* of estimating benefits.

A project may also improve the environment quality, such as better air quality by reducing congestion. Since there are no directly observable market prices for air quality, we need to estimate shadow prices for the benefit. In *contingent valuation* (Mitchell and Carson, 1989), the affected community is surveyed for their willingness to pay (WTP) for the benefit. From a representative sample, it is possible to compute the average annual WTP and then multiply it by the size of the affected population to obtain the aggregate WTP.

However, respondents may not be able to answer hypothetical questions about the environment. A question like "How much are you willing to pay each year for a 10% improvement in air quality?" is not easy to answer. Respondents may also under-declare their WTP to reduce their individual contributions to the cost if the project goes ahead. This is the *free-rider problem* in the provision of public goods. Conversely, respondents may also exaggerate the desirability of a project if they know that the government will provide a substantial subsidy.

A road upgrading project not only reduces transport costs. It may also reduce the number of annual fatalities. The *statistical value of life* approach uses data on the risk of death for various occupations to estimate the benefits of saving human lives. For example, suppose we have the following data:

Occupation	Median annual salary ($)	Deaths per 100,000 workers
A	80,000	50
B	75,000	10
Difference	5,000	40

The value of a life (V) is given by

$$V = \frac{\Delta S}{\Delta P} = \frac{5,000}{40/100,000} = \$12.5m$$

where ΔS is the difference in salary, and ΔP is the difference in probability or risk of death. A weakness of this approach is that it assumes that differences in salaries are due to risk of death. It excludes other factors such as gender, age, education level, and experience.

Non-benefits

Generally, only project outputs are considered benefits, and input-related effects are non-benefits. For example, creating employment by hiring more workers is not a project benefit, even though it is advantageous to the local politician. This is because the money may be spent on other projects that also employ workers. A further reason is that if the economy is at full employment, then the project bids for workers from other sectors. No new employment is created. In a recession, a project may also not generate many new jobs because most firms have excess capacity and do not need to hire new workers. Another reason for the slower hiring is a possible mismatch of skills or lack of experience on the part of new workers. The construction industry tends to lose experienced workers during a recession. When the boom returns after a long period, workers who have left the industry may not return (Guttentag, 1961). Hence, the industry has to constantly train new workers to replace departing skilled workers.

Similarly, providing worker training is also not a project benefit if it is viewed from the input side.

A related non-benefit is the multiplier effect. When workers are hired for a project, they spend their incomes in the local economy, thereby creating secondary effects on other sectors. However, some incomes leak out of the local economy through central government taxes, imports, and the use of their incomes to repay debt. Taken together, these employment and income multiplier effects are probably not large (Bom and Ligthart, 2014; Stupak, 2018).

In summary, project benefits exclude input-related effects such as employment, worker training, and multiplier effects.

Double-counting

Some benefits are excluded because they involve double-counting. For example, an upgraded road may raise property values. It is not a project benefit because the increase in value merely reflects that transport costs have fallen. Hence, to include the rise in property values as well as transport cost savings is to double-count.

Similarly, a new inter-city road that bypasses a town may create new businesses along the road. However, it may also destroy some of the existing businesses in the bypassed town. The net effect is likely to be negligible.

Project costs

The project costs consist of direct input costs and indirect external costs. The direct input costs consist of "hard costs" comprising land, capital, labor, materials, transport costs, and utilities, as well as "soft costs" such as professional fees, statutory approvals, and permits. These inputs are valued at *competitive* prices to reflect the real resource costs. Transfer payments such as profit, interest, subsidies, levies, and taxes do not involve real resources and are excluded from project costs in CBA.

If monopoly, price controls, subsidies, and taxes distort domestic prices for local goods, we need to adjust them to competitive market levels. These adjustments can be complex because of theoretical and data issues. A more practical approach is to use market prices unless the distortions are large. Some projects involve donors, and in-kind donations should be valued at resource cost.

Imported inputs

CBA is normally executed using domestic prices or the local currency, also called the *numeraire* or base prices. In countries where domestic prices do not reflect opportunity costs, it may be possible to use international prices to correct for the distortion. Such price corrections are called *shadow pricing*, and shadow prices are also called efficiency prices. For example, if electricity is heavily subsidized, using the domestically subsidized price will not reflect opportunity costs. In such cases, we may try to use international or border prices (Squire and van de Tak, 1975). However, there is no standard international price for electricity. More generally, international products have differing quality and functionality and, hence, different prices.

These international prices, inclusive of insurance, freight, and local transport costs, are then converted to domestic prices using an appropriate shadow exchange rate (*SER*). Import duties are excluded. Prior to the 1990s, many developing countries had official and unofficial exchange rates that differed widely. Some governments deliberately over-valued their currencies to cheapen the import of equipment and materials for industrialization at the expense of agriculture and other exports. In such cases, the official exchange rate (*OER*) will not be competitive.

Nowadays, the difference has considerably narrowed. For example, since August 2022, the *OER* for the Myanmar kyat has been around 2,100 kyat per US dollar. The unofficial rate is about 2,800 kyat per US dollar. The conversion from *OER* to *SER* is given by

$$SER = k(OER)$$

where k is the conversion factor. An analyst from the public sector may simply use the OER, implying that $k = 1$. Alternatively, it may be argued that *OER* and *SER* are not substantially different. Some analysts use the following trade-weighted formula (Lagman-Martin, 2004):

$$k = 1 + \frac{T}{TT}.$$

Here, T is net trade taxes, and TT is total trade, that is, imports plus exports. If taxes are small relative to total trade, then k is close to 1. Many analysts make this simplifying assumption to obviate the need to estimate k from trade data.

Investment costs

Capital costs include land, site development, facilities, machinery, and vehicles. The investment is booked at a historic cost rather than the commercial practice of depreciating it annually. The commercial practice avoids reflecting a major loss in the first year of operation because of the large initial investment. In CBA, it is unnecessary to spread the initial capital cost over a number of years, which is merely an accounting and tax convention.

Capital investments normally occur during the initial phase of the project. Over time, these long-lived assets require periodic replacement. For example, some machinery may need to be replaced after 10 years. At the end of the project life, these assets have salvage or scrap values. As shown in Table 3.1, scrap values are booked negative because they are benefits, not costs. Further, they are booked in Year 26 (technically, end of Year 25), after the project ends in Year 25. This just an assumption of when the assets are sold. Similarly, some analysts use Year 1 instead of Year 0 as the starting year.

These costs are booked using real prices to avoid having to forecast yearly inflation rates if nominal prices are used. The real price is the nominal price divided by the expected rate of inflation. The inflation rate is expected, implying that it is a forecast and not the actual rate for each year.

Table 3.1   Investment costs ($m).

	Year								
	0	1	2	3	...	10	...	25	26
Land cost	3	0	0	0		0		0	−3
Site development	1	1	1	0		0		0	0
Facilities	20	2	1	0		0		0	−20
Machinery	10	0	0	0		5		0	
Vehicles	1	0	0	0		1		0	
Total	35	3	2	0		6		0	−23

Operating costs

The operating costs for each year of operation consist of fixed and variable costs. Fixed costs are overheads (Table 3.2). Variable costs, such as the amount of raw materials, fuel, and number of workers, vary with the output. In a manufacturing setting, fixed costs are often office costs, and variable costs are factory or site costs. Hence, utilities appear twice as fixed and variable costs.

Table 3.2   Operating costs ($m).

	Year								
	0	1	2	3	...	10	...	25	26
Variable costs									
Materials		0.20	0.40	0.50		0.50		0.50	0
Labor		0.40	0.80	1.00		1.00		1.00	0
Utilities		0.05	0.10	0.20		0.20		0.20	0
Fuel		0.20	0.40	0.60		0.60		0.60	0
Fixed costs									
Management		0.05	0.10	0.20		0.20		0.20	0
Maintenance		0.05	0.10	0.20		0.20		0.20	0
Utilities		0.05	0.10	0.30		0.30		0.30	0
Total		1.00	2.00	3.00		3.00		3.00	0

Recall from the beginning of this section that all inputs in CBA must be competitively priced to reflect actual resource costs. For instance, many governments in developing countries subsidize electricity for political and social reasons. In such cases, the subsidy should be removed to estimate the competitive price. Similarly, if the fuel supplier is a monopoly, we should estimate and use the competitive price and not the monopoly price. One possible way is to use international or border prices. The adjustment is not crucial if utilities make up only a small portion of the total operating cost.

The labor market may not be in equilibrium because of minimum wage legislation, union power, discrimination, and other reasons. If the project hires skilled workers and professionals, the market wage rate, including employment benefits, approximates the opportunity cost of labor.

For unskilled labor, there are two possibilities. If the project uses rural or migrant workers, the opportunity cost is the rural wage rate. If the project employs urban workers, the opportunity cost is the urban wage rate.

For skilled labor, it is necessary to estimate the wage rates for different types and levels of skills.

If the first two years of operation comprise the start-up period, operating costs will rise and stabilize by Year 3. Again, these costs are in real prices to be consistent with investment costs. In Year 26, the facility is no longer in operation, and the values will be zero.

Unlike the income statement of private firms, interest expenses and taxes are not part of the operating cost in CBA. This is because we are interested with resource allocation and not how the investment is financed or taxed. A project should not be deemed feasible because it enjoys generous government subsides, attractive tax breaks, or cheaper financing.

Working capital

In addition to investment and operating costs, a project requires working capital or net working capital. The terminology is not consistent, and we will use the shorter term "working capital" (WC).

In brief, working capital refers to money tied up in a business within a year or the short-term cash flow. Thus

Working capital = Current assets − Current liabilities.

It consists primarily of material and chemical stocks, inventory of goods semi-processed or produced but not yet sold, accounts receivable (AR), and accounts payable (AP). AR refers to goods sold or services rendered by the project company that has yet to be paid, such as if the goods are sold on credit. Similarly, the firm may purchase inputs from suppliers on credit, payable three months later or within a year.

In terms of the flow of costs, it is the *change* in working capital (ΔWC) that matters (Table 3.3). Thus

$$\Delta\text{WC} = \Delta\text{Current assets} - \Delta\text{Current liabilities}.$$

If ΔWC is positive, then ΔCurrent assets > ΔCurrent liabilities. If it is negative, then ΔCurrent assets < ΔCurrent liabilities.

Like investment assets, the remaining stock of materials and goods will be sold in Year 26 after the project company ceases operation.

Table 3.3   Change in working capital ($m).

	Year								
	0	1	2	3	...	10	...	25	26
ΔStocks									
Materials	0	0.5	0.1			0.1		0	−0.1
Goods	0	0.5	0.1			0.1		0	−0.1
ΔAR	0	0.1	0.2			0.1		0	0
ΔAP	0	0.1	0.2			0		0	0
ΔWC	0	1.2	0.6			0.3		0	−0.2

External costs

A project may generate external costs. To estimate such costs, we may use

- loss of output;
- loss of worker earnings;

- replacement or restoration cost;
- averting expenses;
- medical expenses;
- surrogate markets; and
- contingent valuation.

If a firm pollutes a river, it results in the loss of output for fishermen. In transport projects, the loss of earnings may be used to estimate the time cost of congestion. This is called the *human capital approach*. If a project erodes the soil erosion, the external cost is approximated by the cost of restoration.

If a house is next to a noisy road, the owner may install double-glazed windows, which are averting expenses. In pollution studies, one may use the medical expenses of the affected community to estimate the cost of air pollution.

The housing market can act as a *surrogate market* for road noise because buyers will consider the adverse effect on property values. This is called the *revealed preference approach* because actual human market behavior is used to estimate shadow prices. All else being equal, a house facing a noisy road will sell less than a quieter one. This is the principle of *hedonic pricing* (Rosen, 1974), where the price of a house depends on its bundle of characteristics, such as accessibility, land area, built-up area, tenure, age, type of construction, orientation, noise, amenities, and so on. Each characteristic has an unobservable "implicit" price. We can use regression analysis to estimate implicit prices using

$$P_i = f(C, N; \beta), \qquad i = 1, \ldots, n.$$

Here, P_i is the price of the ith house, $f(.)$ is a regression function, C is a vector of house characteristics except noise, N is noise, β is a vector of k coefficients, and n is the sample size. For a linear function,

$$P_i = \beta_1 + \beta_2 A_i + \ldots + \beta_k N_i + \varepsilon_i$$

where A_i is the age of the house and so on, and ε_i is the error term with a zero mean. The expected (mean) value is

$$E(P_i) = \beta_1 + \beta_2 A_i + \ldots + \beta_k N_i.$$

Our interest is the value of β_k. The partial differentiation of $E(.)$ with respect to N gives

$$\frac{\partial E(P_i)}{\partial N_i} = \beta_k.$$

Hence, β_k is the average house price change for each unit change in noise level (in decibels), holding other house characteristics constant. A simpler approach is to make N_i a dummy variable where $N_i = 1$ if a house is noisy and 0 if it is not. Then, β_k is the estimated average price difference between a noisy and a quiet house, holding all other housing characteristics constant.

In practice, we use about 10 to 15 house characteristics such as age, design, land area, built-up area, orientation, number of rooms, accessibility, and so on. Hence, the sample must be reasonably large (e.g. $n = 300$ transacted house sales) to obtain reliable estimates of the coefficients (Tan, 2022).

The housing market must be stable to avoid large price errors, which implies that transacted house sales should be within a short period, such as three months. Further, noise affects buyers of expensive houses more than purchasers of cheaper houses. Hence, β_k varies across income groups and housing markets. Hedonic price estimates of the benefit of noise reduction around airports puts it at about 1% per decibel (Nelson, 1980). Thus, if the desired reduction is five decibels, it will raise nearby property values by about 5%.

The final method of valuing the external cost of a project is *contingent valuation*; that is, we ask affected residents in a survey for their *willingness to accept* (WTA) compensation for the environmental *deterioration*. Recall that we may use a respondent's WTP to value environmental *improvement*. In general, for a given level of improvement or a similar level of deterioration, WTA is greater than WTP because people tend to value a loss more than a similar gain. This proposition of loss aversion comes from the prospect theory of human psychology (Kahneman and Tversky, 1979). The pain of losing $1,000 is more than the gain of the same amount.

Contingent valuation has some weaknesses (Graves, 2014), such as

- differing values across different races, income groups, educational levels, degrees of risk aversion, and so on;
- questionnaire designs that may affect the responses because of emotive reactions towards environmental changes;
- interviewer bias;
- the unwilling of some respondents to accept compensation for destroying nature because it is priceless;
- difficulties dealing with hypothetical "with and without project" or "what if" questions; and
- strategic bias in that respondents may guess how the information may be used and then provide answers towards their preferred outcomes.

Statement of benefits and costs

This annual statement sums up our discussion on the benefits and costs of a project. The primary benefits may consist of output revenues (if the product is marketed), transport cost savings for transport projects, and other secondary benefits. The costs comprise investment and operating costs, as well as the change in working capital. The net benefit is the difference between benefits and costs. The template is shown in Table 3.4.

Table 3.4   Annual benefits and costs ($m).

	Year								
	0	1	2	3	...	10	...	25	26
Benefits									
Investment costs									
Operating costs									
ΔWC									
Net benefits									

Decision criteria

How do we know if a project has improved social welfare? The various criteria are discussed below.

Pareto criterion

In the Pareto criterion, a project improves social welfare if at least one person is better off and no one is worse off. It is impractical because almost every project has winners and losers. It is impossible to find infrastructure projects where there are no losers. For example, in building an airport, some residents will be affected by land acquisition, noise, traffic congestion, and so on.

Kaldor–Hicks criterion

In the Kaldor–Hicks criterion, social welfare has improved if benefits exceed costs, and winners can potentially compensate losers. The compensation is only potential; it may not occur. For example, in a road project, the government is likely to compensate residents for land acquisition but not for noise and dust. Even here, the issue of compensation is often contentious. For example, if farmland is acquired to widen a road, should the compensation value be based on the existing use or the new use if the road is built? The difference may be substantial if a new town is to be built nearby after road widening.

The cost of acquisition may encourage public agencies to acquire land in the poorer areas of the city for public projects. It is sometimes argued that project benefits and costs should be weighted in favor of the poor (Squire and van de Tak, 1975). However, there is no consensus on these distributional weights. Hence, benefits and costs are aggregated without regard to who wins or loses. The rationale is that there are just too many public projects, and it is more efficient to use fiscal policy such as taxation to redistribute income to the poor.

In summary, we add the benefits and costs of a project without regard to who wins or loses.

Net present value

The net present value (*NPV*) criterion is used to rank projects. The benefits (*B_t*) and costs (*C_t*) are discounted to *NPV*, that is,

$$NPV = \sum_t \frac{B_t - C_t}{(1+r)^t}. \tag{3.1}$$

Here, *r* is the *discount rate*, *t* is time, and the summation is over all periods (i.e. annually) from $t = 0$ to $t = n$, the project terminal period (e.g. 25 years). Whether *t* starts from 0 or 1 is a matter of assumption as long as there is consistency. In discounting future benefits and costs, we assume that they occur at the *end* of each period rather than at the start or middle of the period (year).

A project is economically feasible if $NPV > 0$. If it is less than zero, the project is not feasible. It is still possible to make the project feasible by making suitable adjustments, such as by reducing the development costs or thinking of ways to increase the benefits.

More commonly, we rank individual projects using the *NPV* criterion. A project with higher *NPV* is preferred if the projects are mutually exclusive. That is, they are stand-alone projects.

If the public agency has a budget constraint and several projects have positive *NPV*s, it should select the subset of projects that maximizes *NPV*. In Table 3.5, there are four projects, and the present value (*PV*) of their costs ($14 m) exceeds the public agency's budget of $10 m.

Table 3.5   Budget constraint and *NPV* ($m).

Project	PV of costs	NPV
A	2	10
B	3	12
C	4	15
D	5	16
Total	14	

There are several possibilities:

Projects	PV of costs	NPV
A, B, and C	9	37
A, B, and D	10	38
C and D	9	31

The agency should select projects A, B, and D.

Since B_t and C_t are in real terms, the discount rate is also real. The link between real and nominal rates is given by

$$1 + r = \frac{1+R}{1+\pi}$$

where r is the real discount rate, R is the nominal discount rate, and π is the expected rate of inflation. Rearranging,

$$1 + R = (1 + r)(1 + \pi) = 1 + \pi + r + r\pi.$$

For small values of r and π, $r\pi$ is close to 0. Hence,

$$r = R - \pi.$$

That is, the real rate is the nominal rate minus the expected rate of inflation. For example, if the bank quotes a savings deposit rate of 2% and the expected rate of inflation is 1.5%, the real interest rate is 0.5%.

Choice of discount rate

The real discount rate (r) is given by

$$r = W + \lambda.$$

W is the *weighted average cost of capital*, and λ is the *risk premium*. If a project is funded using the following means,

Source of funds	Amount ($m)	Proportion of project cost	Cost of funds (%)
Tax	20	0.20	5
Bond	30	0.30	6
Loan	50	0.50	8
	$100 m	1.00	

then

$$W = 0.20(5\%) + 0.30(6\%) + 0.50(8\%) = 6.8\%.$$

Since r is in real terms, W is also in real terms. W incorporates the normal riskiness of a certain type of project. This is because when lenders lent the $50 m at an 8% interest, they have already considered the project risk. If a project is riskier than normal, it is usual to add a *risk premium* (λ) to the W to reflect the higher risk. Arrow and Lind (1970) argued that $\lambda = 0$ because the public sector has a large portfolio of projects to spread the risk. A counter-argument is that even if this is true, the risks are specific to a project.

Many governments use a real discount rate of about 3–4% for infrastructure projects and slightly lower for $n > 30$ years so that distant benefits and costs do not become negligible. Instead of the government's cost of funds, some analysts use the market interest rates for savings and loans. In a perfect market, the interest rate is determined by the supply and demand for loanable funds. Savers (households) have a positive rate of time preference, implying that the interest rate is positive to entice them to forgo current consumption for a larger share of future consumption. For example, if a saver saves $100, she expects to get $100 + i next year, where i is the real savings interest rate. In actual capital markets, the savings and lending rates differ because of market imperfections. For example, the lender lends higher than the savings rate to cover transaction costs and profit. Different lenders will also offer different savings and lending rates depending on the duration of the deposits and loans, fixed and variable rates, and the credit worthiness of the borrower.

In summary, the discount rate reflects the cost of funds and project risk. There are three possible rates, namely, the public sector cost of funds, the community savings rate, and the private sector borrowing rate. It is also possible to use a weighted average of all three rates.

Internal rate of return

Instead of searching for a discount rate to use in the NPV formula, we may compute the project internal rate of return (IRR) k by setting $NPV = 0$ and solving for k:

$$0 = \sum_t \frac{B_t - C_t}{(1+k)^t}. \tag{3.2}$$

Suppose a project has the following benefits and costs:

t	Benefit ($m)	Cost ($m)	$B_t - C_t$
0	0	100	−100
1	40	10	30
2	50	10	40
3	60	10	50

Then, using a 5% discount rate,

$$NPV = \sum_t \frac{B_t - C_t}{(1+r)^t} = -100 + \frac{30}{1.05} + \frac{40}{1.05^2} + \frac{50}{1.05^3} = \$8.1m. \tag{3.3}$$

To find the project *IRR*, we solve for k:

$$0 = \sum_t \frac{B_t - C_t}{(1+k)^t} = -100 + \frac{30}{1+k} + \frac{40}{(1+k)^2} + \frac{50}{(1+k)^3}.$$

This is done by varying k until the value on the right-hand side (RHS) is close to 0:

K	RHS
0.08	0.18
0.09	−0.02
0.089	0

Hence, the project *IRR* is 8.9%, which is higher than the 5% discount rate. In practice, it is easier to use software to compute the *IRR*. For example, in Microsoft Excel, we enter the function "=IRR(x:y)" (without the quotes), where x is the first cell and y is the last cell of the column of net benefits.

It is possible to have multiple roots in the *IRR* equation if there are alternating cash flows. However, this is rare for infrastructure projects. Generally, the project incurs costs during the pre-construction and construction stages and generates benefits during operation.

The project *IRR* is compared with the *hurdle rate*, the government's minimum required rate of return. In many cases, the hurdle rate is the weighted average cost of capital.

Like the *NPV* criterion, we may also use *IRR* to rank projects. Projects with higher *IRR* are preferred.

The *NPV* and project *IRR* criteria may not be consistent. In Fig. 3.4, we plot two *NPV* against r "curves" using Equation (3.3), and the "curves" are approximated by straight lines for ease of exposition. At a lower rate of discount r, project *B* is preferred because the *NPV* is higher. By the *IRR* criterion, project *A* is preferred because k_2 is greater than k_1.

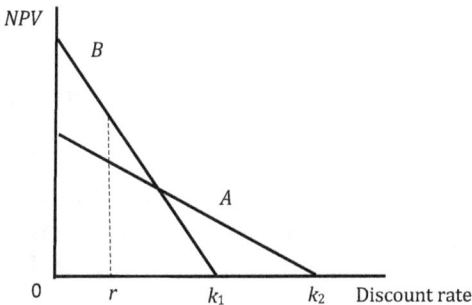

Fig. 3.4 Inconsistency of *NPV* and *IRR* criteria.

Cost-benefit analysis of a proposed road project

We now consider a simple example of CBA to put all the elements together (Table 3.6). A proposed road-upgrading project has the following costs and benefits ($m):

Here,

$$\text{Net benefit} = \text{Benefits} - \text{Costs}.$$

Table 3.6   Costs and benefits for the road project.

	Year			
	0	1–4	5–29	30
Benefits	0	0	3.2	3.2
Costs				
Land	100	0	0	−100
Construction	0	50	0	0
Operation and Maintenance	0	0	1	1
Noise	0	0	1	1
Net benefit	−100	−50	1.2	101.2

The benefits include

- annual travel cost savings computed using Fig. 3.3;
- annual fall in road fatalities, estimated using the statistical value of life approach; and
- improvement in air quality, estimated using contingent valuation.

The construction period covers Years 1 to 4, and the column figures are construction expenses ($50 m) for a typical year of construction. The operation phase is from Years 5 to 30, and the column figures are for a typical year to avoid too many columns.

By using real benefits and costs, there is no need to adjust the values for annual inflation. In the terminal year (Year 30), the land reverts, and its value is booked as negative (−100). Hence,

$$NPV = -100 - \frac{50}{1+r} + \cdots + \frac{1.2}{(1+r)^5} + \cdots + \frac{101.2}{(1+r)^{30}}$$

The project *IRR* is found by setting $NPV = 0$ and replacing r with k.

Risk assessment

There are many ways to assess project risks, including

- sensitivity analysis;
- Monte Carlo simulation (MCS);

- risk-adjusted discount rate;
- project options; and
- qualitative method.

The risk-adjusted discount rate method has been discussed under the choice of a discount rate. The basic idea is to include a risk premium to the weighted average cost of capital. The other methods are discussed below.

Sensitivity analysis

In sensitivity analysis, we investigate how *NPV* changes if we alter *one* variable and hold other variables constant. The most popular choice is to change the discount rate, such as using 3%, 6%, and 9%, respectively, and see how *NPV* varies.

It is also possible to vary the forecast revenues or a major cost item and see how *NPV* changes. In practice, many analysts use MCS because it examines the effect of *several* variables on *NPV*.

Monte Carlo simulation

In MCS, we use random numbers to change the variables *simultaneously* to see how *NPV* changes. The popular choices are changes in the discount rate, revenue, and initial investment cost (Table 3.7). The second and third columns show the possible values of each variable and the subjective probability distribution, respectively. For example, the discount rates are assumed to be 5%, 7%, and 9%, respectively. These values and probabilities are derived from experience. For example, there is a 0.2 chance that the annual revenue may turn out to be $45 m, a 0.7 chance that it will be $40 m, and a 0.1 chance that it will be $35 m. The next column gives the cumulative probabilities based on the probability distribution in the previous column, which must sum to unity. Finally, we assign corresponding numbers between 00 and 99 based on the cumulative probabilities. It is possible to assign 4-digit numbers (0 to 9,999), but for simplicity, we will use 2-digit numbers.

Table 3.7   Data for the Monte Carlo simulation.

Variables	Possible values	Probability distribution	Cumulative probability	Assigned numbers
Discount rate (%)	5	0.3	0.3	0–29
	7	0.5	0.8	30–79
	9	0.2	1.0	80–99
Revenue ($m)	45	0.2	0.2	0–19
	40	0.7	0.9	20–89
	35	0.1	1.0	90–99
Initial cost ($m)	95	0.2	0.2	0–19
	100	0.6	0.8	20–79
	105	0.2	1.0	80–99

In each trial, we draw three random numbers between 00 and 99, similar to the casinos in Monte Carlo. For example, if the first trial draws the set (25, 65, 10), it corresponds to (5%, $40 m, $95 m) in the table. We use these three values to compute the first *NPV*, assuming that the revenue of $40 m is the same for each year. If the second trial draws the set (80, 4, 35), it corresponds to (9%, $45 m, $100 m). We use these values to compute the second *NPV*, and so on.

By using a large number of trials, we can develop frequency tables and use them to compute the mean and standard deviation of *NPV*. The latter is a measure of project risk from simultaneous changes in the three variables. If desired, other variables, such as input costs, may be included.

There are many software on MCS, such as the @Risk add-in function in Microsoft Excel. Some versions allow correlation among the variables, which is more realistic than assuming independence. For example, if the project is completed during a recession, the revenues for the initial years may be low, and we may want to use a lower discount rate instead of a randomly selected one.

Project options

So far, we have assumed that a project is evaluated as a whole; that is, it is either approved or not approved. This is the case if it is a bridge, where it is not possible to build half a bridge. In projects such as a high-speed rail (HSR), it is possible to develop the project in phases. For example, if the entire HSR has 15 stations, it is possible to start with 5 stations in phase 1, another 5 stations in phase 2, and the last 5 stations in phase 3. These are *real options* because they use up actual resources, as distinct from financial options given to senior management of listed companies.

Similarly, a private developer may break up a large development into different phases. If the first phase is successful, the developer will launch the second phase. However, if the first phase is not successful, the developer has the option to continue with the project (because the economy has picked up), make changes to the design to make the development more attractive, delay the launch, or abandon the project. By executing the development in phases, the developer avoids the risk of a major loss and is able to learn from the experience in earlier phases.

Suppose a project has three periods, Year 0, Year 1, and Year 2 (Fig. 3.5). The boxes represent *NPV*s (in $m). In Year 1, the project has a 0.8 chance of earning $70 m and a 0.2 chance of losing $10 m by being decommissioned. The *NPV* for Year 0 is $0.8(70) + 0.2(-10) = \$54$ m. In Year 1, the *NPV* is $0.6(90) + 0.4(40) = \$70$ m. Hence, we work backwards to compute the *NPV* for each phase. At each point, the developer will need to compute the *NPV* based on different scenarios. Since *NPV* = $54 m, the project is worth pursuing in the initial phase. However, if it turns out bad in the next phase, the developer may abandon the project.

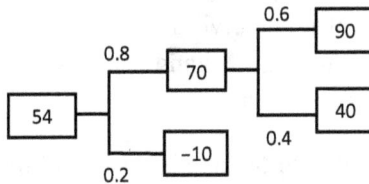

Fig. 3.5   Project with real options.

Qualitative method

The most common technique of assessing project risks is the qualitative method. This is because quantitative risk assessments tend to focus on financial variables and ignore many other risks. The qualitative method is more comprehensive and provides the opportunity for each party in the contract to structure or restructure the project to manage the risks. This method is used extensively in this book.

The *commercial risks* of a project include

- *completion risks*, where the project fails to achieve the construction time, cost, safety, and quality targets;
- *market risks*, where revenue is insufficient because of price or volume, high input costs, or non-payment by buyers;
- *operation and maintenance risks* relating to equipment failure, inefficient operation, and input supply issues;
- *financial risks* relating to movements in foreign exchange, inflation, and interest rates;
- *social and environmental risks* relating to heritage, resettlement, illegal or unhealthy work practices, perceived lack of participation by affected parties, and environmental damage; and
- *force majeure risks* such as war, civil unrest, general strikes, epidemics, pandemics, and natural disasters that directly destroy the project assets or indirectly disrupt its operations.

The *non-commercial risks* include

- *political risks* through acts of government such as a change in government, nationalization, expropriation, issues with permits and approvals, inability to repatriate profit, restrictions on operation, and general instabilities such as war and civil unrest; and
- *regulatory risks*, including a change in law, protection of property rights, and dispute resolution.

There appears to be some overlap between political and *force majeure* risks. However, unlike *force majeure* events, political risks refer to events that arise through deliberate government action.

Risk assessment involves ascertaining the likelihood (L) of a risk event and its impact (I). Risk exposure is the product $L \times I$, as shown in the risk assessment matrix for non-commercial risks in Table 3.8. The scoring is based on a 1–5 rating scale, with higher scores representing greater likelihood and impact.

Table 3.8   Risk assessment matrix for non-commercial risks.

Risk	Likelihood (L)	Impact (I)	Exposure (L × I)	Mitigation
Political				• Political risk insurance
Tax changes	1	1	1	• Implementation Agreement (if possible)
Social opposition	3	4	12	
Terrorism	1	4	4	• Involve local sponsors and multilateral development agencies in financing
Expropriation	1	5	5	
Repatriation of profit	1	5	5	
Production restrictions	1	5	5	
Trade restrictions	1	5	5	
Corruption	3	4	12	
Regulatory				• Develop a PPP framework, seek legal advice, contractual allocation of risks
Regulatory framework	1	3	3	
Permits	1	3	3	
Change in regulation	2	3	6	
Legal framework	1	3	3	
Disputes	5	4	20	
Changes in law	1	1	1	
Environmental				• Resettlement plan
Opposition	3	4	12	• Community participation and inclusion
Damage	1	3	3	• Environmental Impact Assessment and mitigation
				• Insurance against clean-up cost
Force majeure	1	5	5	• Insurance
				• *Force majeure* provisions in a PPP contract
Land acquisition	4	5	20	• Land Acquisition Act
				• Stakeholder management
				• Fair compensation
End of concession	5	3	15	• Link final payment to asset condition in a PPP contract

In general, risk is allocated to the party that is best able to manage it. Hence, the grantor assumes the non-commercial risks while the SPV manages the commercial risks. If both parties have no control over the risk event, such as a natural disaster, they should share the risk.

A party may mitigate project risks by

- avoiding it, such as not to initiate or bid for the project at the expense of future benefits or profits;
- transferring it to another party, through contractual risk allocation or insurance;
- retaining it, if the risk is manageable; and
- reducing it, through contracts, hedging, and other means.

Many commercial risks arise from *imperfect* or *asymmetric information* about the future, product quality, work effort, and contractual behavior (Table 3.9). Information is imperfect if it is not fully known. For example, if we do not know what the future will bring, we will say that there is a hidden future. Asymmetric information refers to the uneven distribution of information between two contracting parties. For example, a shoddy contractor may try to hide product defects from the client. This is the problem of hidden information. Similarly, the borrower knows more about her credit worthiness than the lender, and may hide this information to secure a loan.

The problem of the hidden effort is everywhere. How does the superior ensure that the subordinate will put in the work effort? Finally, hidden

Table 3.9   Risks from imperfect information and mitigation.

Imperfect information about	Mitigation
Future revenues	Forecast models, hedging instruments, insurance
Product defects	Specifications, screen, test, measure, external review, track record, warranties, certification, and advertising
Effort	Incentives, penalties, supervision, appeal to values (culture), and indoctrination
Action	Excess and caps (in insurance), vertical integration, long-term fixed contracts, diversify sources

action refers to post-contractual *opportunism*. For example, a car driver may become less careful because damage to the car is insured. For this reason, insurers impose excess and caps in their policies. An excess means that the driver will pay for minor damages below a certain sum, e.g. $1,000. A cap is the maximum amount the insurer will pay if the car is damaged. If a firm finds that the supplier may behave opportunistically, it may vertically integrate the process; that is, it will not outsource it. Finally, to prevent suppliers from unreasonably jacking up prices, a firm may diversify its sources of supply.

During the implementation stage, there is a need to monitor the risks as they may change or new risks may appear. Periodically, there should be a major risk review after each project milestone.

Recommendation

The final step in CBA is to recommend whether the project is economically feasible. Note that the analysis is purely economic, and the final approval will need to consider technical, financial, political, and other considerations.

Criticisms of cost-benefit analysis

There are many criticisms of CBA (Graves, 2007). They include

- political and stakeholder interference to influence the outcome (Pressman and Wildavsky, 1979);
- difficulties in valuing benefits and costs, particularly those involving shadow prices where data is harder to collect, and the willingness to pay for a good or service as well as the willingness to accept compensation for a deterioration of the environment is dependent on income;
- the exclusion of certain stakeholders, such as the poor, future generations, and those who are adversely affected or displaced by the project from the process (Mishra and Prasad, 2020);
- differing opinions on the choice of discount rate;
- distributional issues, particularly the impact of a project on the poor;
- non-compensation of losers or the compensation is inadequate, especially with regard to contentious land acquisition (Sathe, 2017); and

- the use of market values often results in public acquisition of cheaper land from poorer areas of the city, resulting in the displacement of working-class neighborhoods (O'Conner, 1995).

Despite these and other criticisms, CBA is still widely used for the evaluation of public projects for the lack of a suitable alternative.

Political, regulatory, and legal assessment

A project may generate political support or opposition. Those who oppose it are likely to cite land acquisition, environmental concerns, user afford-ability, excessive costs, and insufficient benefits. The project will also need to meet regulatory requirements and be assessed for potential legal issues, particularly in countries with weak regulatory frameworks and legal systems.

For PPP projects, an assessment of bankability is required. A project is bankable if it can attract commercial lending. It may be possible to provide some government support or change the terms of the PPP con-tract, such as to lengthen the concession period to improve bankability.

Business case

The business or policy case sums up the preliminary study.

Traditionally, a project is considered "successful" if it is delivered to specifications, completed on time, and is within budget. However, it may still be a white elephant. Hence, it is necessary to provide the business or policy case, which considers the value of the project to stakeholders (Thoradeniya and Tan, 2018).

For commercial projects, the business case is likely to be

- profit;
- process improvement;
- strategic; or
- compliance with regulatory requirements.

The strategy may be an expansion into a new market, launch of a new product, or investment to deter the entry of competitors.

For public projects, the high-level policy case at this stage may be social welfare improvement or value for money, as discussed earlier under the section on technical assessment. The contents of the policy case include

- background;
- project definition;
- project governance structure;
- technical feasibility;
- economic feasibility;
- financials;
- project delivery;
- procurement; and
- risk management.

Observe that the policy case contains the key points of a preliminary study.

References

Arrow, K. and Lind, R. (1970) Uncertainty and the evaluation of public investments. *American Economic Review*, **60**(3), 364–78.

Bom, P. and Ligthart, J. (2014) What have we learned from three decades of research on the productivity of public capital? *Journal of Economic Surveys*, **28**(5), 889–916.

Derman, W. and Whiteford, S. (2019) *Social impact analysis and development planning in the Third World*. London: Routledge.

Goodwin, P., Dargay, J. and Hanly, M. (2004) Elasticities of road traffic and fuel consumption with respect to price and income: A review. *Transport Reviews*, **24**(3), 275–92.

Graves, P. (2007) *Environmental economics: A critique of cost-benefit analysis*. New Delhi: Rawat.

Graves, P. (2014) *Environmental economics*. London: CRC Press.

Griffiths, R. (2017) *Revitalizing the silk road*. Leiden: HIPE Publications.

Guttentag, J. (1961) The short cycle in residential construction, 1946–59. *American Economic Review*, **LI** (3), 278–98.

Hayek, F. (1976) *The road to serfdom*. Chicago: Chicago University Press.

Hebbar, K. (1991) *Integrated rural development program*. New Delhi: Deep and Deep.

Kahneman, D. and Tversky, A. (1979) Prospect theory: An analysis of decision under risk. *Econometrica*, **47**(2), 263–91.

Lagman-Martin, A. (2004) *Shadow exchange rates for project analysis*. Manila: Asian Development Bank.

Levy, J. (1995) *Essential microeconomics for public policy analysis*. New York: Prager.

Mishra, S. and Prasad, S. (Eds.) (2020) *Displacement, impoverishment, and exclusion*. London: Routledge.

Mitchell, R. and Carson, R. (1989) *Using surveys to value public goods: The contingent valuation method*. New York: Resources for the Future.

Nelson, J. (1980) Airports and property values: A survey of recent evidence. *Journal of Transport Economics and Policy*, **14**(1), 37–52.

O'Conner, T. (1995) *Building a new Boston: Politics and urban renewal, 1950–70*. Boston: Northeastern University Press.

Pressman, J. and Wildavsky, A. (1979) *Implementation*. Berkeley: University of California Press.

Rondinelli, D. (1993) *Development projects as policy experiments*. London: Routledge.

Rosen, S. (1974) Hedonic prices and implicit markets: Product differentiation in pure competition. *Journal of Political Economy*, **82**(1), 34–55.

Sathe, D. (2017) *The political economy of land acquisition in India*. London: Palgrave Macmillan.

Shu, F. and Hyndman, R. (2011) The price elasticity of electricity demand in South Australia. *Energy Policy*, **39**(6), 3709–19.

Squire, L. and van de Tak, H. (1975) *Economic analysis of projects*. Washington DC: World Bank.

Stupak, J. (2018) *Economic impact of infrastructure investment*. Washington DC: Congressional Research Service 7–5700, R44896.

Tan, W. (2022) *Research methods: A practical guide for students and researchers*. Singapore: World Scientific Publishing.

Thoradeniya, D. and Tan, W. (2018) Strategic value of a Chinese-funded infrastructure project in Sri Lanka. *Infrastructure Asset Management*, **7**(2), 127–33.

World Bank (2019) *Policy guidelines for managing unsolicited proposals in infrastructure projects*. Washington DC: World Bank.

Chapter 4

Feasibility study

Project Brief

If the high-level business case developed from the preliminary study is approved, the next step in the infrastructure development cycle is to proceed to a more detailed feasibility study.

From the preliminary study, the grantor develops the Project Brief as the basis for briefing the project team before embarking on the feasibility study. The Project Brief is sometimes called the Project Initiation Document, Project Charter, or Statement of Work, depending on the organization.

It may be necessary to appoint external consultants to assist with the feasibility study. If new issues surface during the preliminary study, the grantor will also gather additional inputs from the stakeholders.

The Project Brief contains the following information:

- project name and date;
- project authorization;
- project manager;
- background;
- basic development options;
- preferred option;
- project benefits;
- project scope: output and basic specifications;
- site, including constraints and opportunities;
- basic engineering and design options;
- sustainability and resilient considerations;
- preliminary schedule and milestones;
- revenues and costs;

- stakeholders;
- socio-environmental impacts and mitigation strategies;
- choice of delivery;
- procurement strategy;
- project governance structure;
- regulatory requirements; and
- management of risks.

The list is not exhaustive. It can be tailored to suit different types of infrastructure projects.

Feasibility study

For public–private partnership (PPP) projects, it is the special purpose vehicle (SPV) that finances, designs, builds, operates, and maintains the facility. Hence, the grantor's feasibility study covers the following areas:

- output and technology specifications;
- site analysis;
- site appraisal;
- project schedule, milestones, and length of concession period;
- development of the shadow bid;
- political feasibility and user affordability;
- fiscal sustainability;
- regulatory and legal assessment;
- economic feasibility;
- socio-environmental impacts and mitigation strategies;
- PPP contract structure;
- project governance structure;
- bankability; and
- possible government support.

We discuss these elements below. Many of these items have been considered in the preliminary study. At this stage, it is necessary to confirm some issues, such as economic feasibility, and go deeper into others, such as site analysis.

Output and technology

The grantor needs to specify the output and technology without over-constraining how the SPV may deliver the product or service. For example, in the case of power generation, the grantor may specify the output capacity, operational availability, heat rate, energy sources, and transmission requirements. For a hydroelectric power plant, its location, type (impoundment, diversion, or pumped storage), and turbine type will be specified. For a school, the grantor will specify the functional areas such as classrooms, toilets, lecture rooms, laboratories, canteen, hall, learning garden, parking, security, assembly area, staff and general offices, library, and sports facilities.

It is prudent to build in some technological or output flexibility because of the long concession period. For example, a facility may contain "white" spaces for future use, areas for physical expansion, convertible spaces, and possible changes from non-renewable to renewable technologies.

There is a preference among lenders and grantors for tried and tested technology to minimize the risk of failure. In some cases, grantors deliberately specify the use of a new technology for political, strategic, or commercial reasons.

Site analysis

For infrastructure projects that are procured using PPP, the grantor normally acquires the site and leases it to the SPV, usually for a nominal sum, for the duration of the concession period. Alternatively, the grantor may arrange for the SPV to lease it from another public entity, such as the land authority.

Site analysis refers to assessments of the legal, physical, built-up, external, and social aspects of the site. These factors affect the value of a site and, hence, the cost of land acquisition.

The legal aspects of a site include the boundary, ownership, title, tenure, covenants, regulatory requirements, property tax obligations, performance obligations, and possible development charge or fee. Land titles may include covenants to preserve the character of the neighborhood, such as the permissible use of certain designs or types of construction materials. The planning authority may impose performance obligations as part of town planning approval, such as the need to provide a park or walkway. Finally, a development charge (or fee) is applicable if there are changes to permissible land use or density of development. The charge is levied as a percentage of the change in land value to encourage redevelopment of the site.

The physical characteristics of a site include its size, shape, topography, drainage, vegetation, geology, and soil. These features present constraints and possibilities, including the size of the project, scenic views, sun shading, potential flooding, type of foundation, and possible land contamination.

The built-up features include improvements such as existing buildings, utilities, heritage or conserved structures, roads, and other structures.

A site has many external features, such as

- accessibility to major centers and transport modes;
- surrounding land uses and amenities, such as schools, parks, and markets;
- undesirable features, such as being next to a slum; and
- environmental qualities, such as traffic, noise, wind, air, snow, and sun.

Depending on the type of project, it may be necessary to consider the ease of movement during construction as well as for project outputs and inputs, such as materials, fuel, labor, plant, and equipment.

The social aspects of a site concern stakeholders, neighbors, land acquisition, possible resettlement, employment changes, possible environment damage, traffic, and other community issues.

Following a site analysis, the grantor may consider land use options and configurations, including the development of a site master plan. If it is a small site, this task may be left to the private sector.

Site appraisal

Site appraisal is necessary to determine the rent of the site or cost of land acquisition. The government may acquire a single site or many sites along a proposed railway line or highway. These sites may be appraised using comparable property sales data in the vicinity and checked using rental data. The latter is called the income approach to value, that is,

$$V = \frac{R}{k}$$

where V is the capital value of the asset (e.g. a house), R is the current annual gross rent, and k is the gross capitalization rate (e.g. 4%). If the net rent is used, then k is the net capitalization rate. To "capitalize" means to convert the rent into capital or asset value, in this case, by dividing R by k. We often use the gross rent because the information is more readily available than net rent. The net rent is computed by deducting maintenance, repairs, property tax, insurance, and other expenses from the gross rent. The value of k is obtained from comparable sales data; that is, by computing R/V for similar properties preferably, but not necessarily, in the neighborhood. The assumption is that the rate of return is competitive across different asset classes after adjusting for risks. For example, investing in office buildings should yield similar returns after adjusting for site and other differences.

For private development projects, the developer will need to purchase or lease the land if he or she is not the owner. The value of the land (L) can be estimated using comparable land sales, but because development projects tend to be large and unique, there may not be comparable land sales in the vicinity. A common approach is to compute L as a residual by deducting the initial building cost (C), interest expense during construction (I), sales and letting costs (S), and developer's profit (π) from the capitalized revenue, or gross development value (GDV). For example, for an office building,

$$L = GDV - C - I - S - \pi.$$

Here

$$GDV = \frac{R}{k}$$

where R is the current annual gross rental income and k is the market capitalization rate. This is called the income approach to value. The building cost consists of the hard costs of site preparation and construction, as well as the soft costs comprising professional, planning approval, funding, legal, and other fees (Cadman and Topping, 1995).

Example: Property development

In this example, we compute the average selling price for a private housing development given the following data and assumptions:

Site area $A = 50,000$ ft^2
Plot ratio $\rho = 2.0$
Building design efficiency $= \eta$
Gross floor area $GFA = \rho A = 100,000$ ft^2
Net floor area $NFA = \eta GFA = 0.9(100,000) = 90,000$ ft^2
Profit rate $= \pi = 0.20$
Average unit size $S = 1,000$ ft^2
Number of units $= NFA/S = 90$

The plot ratio is the site density, or GFA/A. Observe that the unit for land area is ft^2 and not m^2 because land is expensive and is traditionally quoted in ft^2. The building design efficiency is the ratio of NFA to GFA. The NFA is the saleable part of a building; that is, it excludes common areas such as corridors, stairs, and lift landings.

The total development cost is given below:

	$m
Land	80
Construction	30
Land financing	10
Fees	5
Taxes	5
Marketing, etc.	10
Total	140

Breakeven price $= \$140$ m$/NFA = \$140$ m$/90,000 = \$1,556$ ft^2
Average selling price $= (1+\pi)1,556 = 1.2(1,556) = 1,867$ ft^2

The fees are for design and other consultants, town planning, legal, financing, and other requirements. These figures are only estimates. Property development is risky (McNellis, 2016), and property cycles can be volatile (Grenadier, 1995).

Before a private developer purchases a potential site, he normally buys an option from the seller to keep it from the market. The option premium is about 1% of the agreed purchase price, and the period is about three to six months for the developer to conduct a feasibility study, secure financing, and obtain in-principle town planning approval. If the project is not feasible, the option expires worthless, and the developer loses the premium. If the project looks promising, the premium becomes part of the down payment for the purchase of land. Other methods of acquiring land for private development include government land sales, joint development with the landowner, and using a long-term ground lease.

In summary, the grantor may need to acquire land for the project. The land value is normally appraised using market sales data of comparable properties. If a large tract of land is required, it may be necessary to use the residual method to compute the land value. Finally, it is useful to know how a developer carries out the financial analysis for a project.

Project schedule, milestones, and length of concession period

At the feasibility stage, the grantor's construction schedule is approximate and largely based on past experience. It is usual to break up a large project into different phases as project milestones.

The length of the concession period should allow the SPV to make an adequate return based on the shadow bid model. It is possible to use a variable concession period based on target revenue, net present value, or return on investment. The SPV may seek to renegotiate the length of the concession period if there are unpredicted risks that lead to a substantial fall in revenue, such as the COVID-19 pandemic.

Shadow bid model

The shadow bid model serves as a guide for the grantor to compare bids. Generally, these bids will need to cover

- debt service and return to equity;
- fixed operation and maintenance expenses; and
- variable costs such as fuel and electricity.

There may be other payments, such as electricity generated in waste-to-energy (WTE) plants. For some projects, such as sports facilities, the grantor and SPV may share the revenue. The grantor needs to estimate the bidder's Income Statement to develop the shadow bid. This is discussed in Chapter 7.

Usually, there are periodic reviews of tariffs and fees to adjust for demand changes, inflation, fluctuations in input prices, and currency movements. For energy-intensive projects such as electricity generation and desalination, there is often a pass-through mechanism to shift the risk of fluctuating input prices from the SPV to the user or off-taker.

For projects in developing countries, the tariff may allow for currency adjustments if the SPV borrows in a stable foreign currency (e.g. US$) but revenues are in a less stable local currency.

Political feasibility and user affordability

Political feasibility and user affordability have already been covered during the preliminary study. These issues will be revisited if there are major changes, such as the emergence of new stakeholders.

Fiscal sustainability

Fiscal sustainability refers to the grantor's ability to make periodic payments under the PPP contract until the end of the concession period. If the project charges user fees and there is a public subsidy, the government must be able to continue to provide the subsidy and compensate users for the higher fees if there are subsequent tariff adjustments.

Governments face various types of fiscal risks (Budina *et al.*, 2007). Macroeconomic risks are related to business fluctuations that may spike the public debt because of issues, such as the fall in tax revenues, Keynesian stimulus and unemployment spending, and higher interest expenses.

Governments may have contingent liabilities such as guarantees on loans for state-owned enterprises and lower levels of governments and public liabilities for other PPP projects. There are also risks of natural disasters that may spike public spending. Finally, there are institutional fiscal risks such as corruption and the absence of internal controls to curb excessive spending.

Regulatory and legal assessment

The grantor needs to assess the clarity of legal contract clauses, regulatory design and quality, extent of public liability, and enforcement of contracts. Of importance is the need to streamline complex procedures, facilitate the approval of permits and approvals, and address conflicting legislation. Clearly, the grantor needs to seek legal and other opinions on many of these issues.

Socio-environmental impacts and mitigation strategies

At the feasibility study stage, it is necessary to conduct in-depth studies on social concerns such as land acquisition and resettlement. The environmental concerns include the sustainability of resources, pollution, conservation of culture and heritage, and loss of habitat and biodiversity.

There are two phases in an environmental impact study. Phase I is common for infrastructure projects, and if major issues crop up, there is a more detailed Phase II investigation and proposed mitigation measures.

In terms of ecologically sustainable development, the possible adverse impacts include (O'Hara, 2014)

- removal of trees;
- destruction of the soil structure and soil contamination;
- increase in rainfall run-off because of hard surfaces;
- barriers to the use of adjoining land by the community because of disamenities;
- air, water, land, light, and noise pollution;
- hazardous wastes;

- loss of habitats;
- degradation of the surrounding environment;
- loss of biodiversity;
- obstruction to the movement of animals;
- pressure on water supplies and other infrastructure (Hedberg, 2020);
- reduction in urban resilience, such as to floods (Coaffee and Lee, 2016); and
- pollution from the use of non-renewable energy sources.

There may be positive impacts from infrastructure development, and these impacts may arise by reversing the negative impacts above. For example, it may replant rather than remove trees.

The mitigation of these adverse impacts requires an effective policy, regulatory, and tax framework. The policy framework requires the adoption of appropriate social and environmental policies. The regulatory framework requires the effective exercise of planning and development, as well as environmental assessments and controls, when approving projects. In particular, projects should incorporate sustainable and resilient designs (see Chapter 1) and adopt green construction materials and methods. Finally, governments may use tax and other incentives to encourage sustainable and resilient practices.

Mitigation requires ongoing evaluation of policy, regulation, and tax regimes. The social and environmental impacts will need to be closely monitored so that remedial actions may be undertaken.

Public–private partnership contract structure

The basic PPP contract structure is Build-Operate-Transfer (BOT). There are variations, such as Build-Operate-Own (BOO), if the asset has little residual value to transfer at the end of the concession period.

In a Build-Transfer-Operate (BTO) project, the transfer of asset takes place after the construction phase, and the grantor then gives the SPV the right to operate the facility during the concession period. In general, in PPP contract structures, the government owns the asset and gives the SPV the right of operation. In some cases, such as in BOO projects, the SPV owns the asset.

The risk allocation follows the principles discussed in Chapter 3. The grantor manages non-commercial risks, and the SPV handles commercial risks. Both parties should share risks beyond their control.

Project governance structure

The project governance structure spells out the decision-making authority, that is, who decides in the project. A large infrastructure project may involve different ministries; hence, there is a need to set up a steering committee comprising senior bureaucrats to oversee the project. Often, the senior official from the Ministry of Finance or primary ministry heads this committee.

The term *governmentality* refers to the way the governance team present themselves as well as how they shape and guide the conduct of stakeholders (Foucault, 2010). Foucault focused on the signal of frugal government and the discourse of collaborative "partnerships."

Bankability

A project is bankable if it is able to attract private lending. The grantor's assessment of bankability considers whether the project makes commercial sense to entice sponsors to bid for it.

If the project is not bankable, the grantor may consider changing the terms of the contract and possible government support to make it bankable. For example, it may consider lengthening the concession period.

Possible government support

The possible government support includes (Irwin, 2003)

- capital grant or subsidy;
- tax incentives;
- equity participation;
- guarantees; and
- building of connecting and ancillary facilities.

For example, the government may provide a guarantee against the default of a public entity as an off-taker or input supplier. The support may also be non-financial, such as helping to win community support for the project (Stein, 1992).

Market sounding

As part of the feasibility study, the grantor sounds out the market about six months to a year before the tender (depending on the complexity of the project) to gather feedback and gauge private sector interest.

Market sounding should not be too early when the project is still in its early stage of design because the feedback will be vague. Conversely, it should not be too late, such as when the design is already fixed.

The feedback should be from industry networks and not from particular firms to avoid perceptions of partiality. The grantor gathers information about output specifications, market conditions, technologies and costs, payment mechanism, and risk allocation from the key stakeholders.

The feedback may take the form of written comments, survey responses, or interviews based on the circulated project information memorandum.

Example: Market sounding for a WTE project in the Maldives (January, 2019)

This open market sounding exercise by the Ministry of Environment (MOE) sought to obtain public feedback. In 2016 and 2017, MOE completed the concept design and feasibility study for a regional waste management facility (RWMF) in Thilafushi.

The key stakeholders are the Ministry of Finance (executing agency), MOE (implementing agency), a government-owned firm (Waste Management Corporation Ltd) that currently runs the waste collection, the State Electricity Company (potential off-taker), the Environmental Protection Agency, and Greater Malé Industrial Zone Ltd (GMIZL), the regional industrial development entity.

The RWMF will consist of a harbor and reception area for the reception of waste, a construction and demolition waste (CDW) processing plant, an end-of-life vehicle dismantling unit, a WTE plant, a bottom ash processing plant, leachate storage and treatment facilities, and a seawater inlet and outlet structure. MOE will conduct studies to determine the demand for CDW, bottom ash, and recyclables.

The market sounding notice provided information on the site and its constraints, type of contract (Design, Build, and Operate (DBO)), contract form and risk allocation (FIDIC DBO contract), main characteristics of the WTE plant, anticipated waste composition, qualification criteria for bidders, and broad project schedule. The proposed payment terms and mechanism include monthly fixed and variable fees, asset replacement reimbursement according to the contractor's bid price schedule, and adjustment for inflation and exchange rate movements. Both parties will share the revenue from the sale of electricity.

MOE is particularly interested in obtaining feedback on the geotechnical and environmental risks of building the WTE plant on a newly reclaimed island on a coral reef.

Business case

Recall from the previous chapter that a high-level business case follows from a preliminary study to determine if the project should proceed to the feasibility study stage. Likewise, a business case is required after the feasibility study to seek approval on whether the project should go ahead.

The format of the business case follows the items in the feasibility study with clear justifications in support of the project.

References

Budina, N., Brixi, H., and Irwin, T. (2007) *Public-private partnerships in the new EU member states: Managing fiscal risks*. Washington DC: World Bank.

Cadman, D. and Topping, R. (1995) *Property development*. London: Routledge.

Coaffee, J. and Lee, P. (2016) *Urban resilience*. New York: Springer.

Foucault, M. (2010) *The birth of biopolitics*. London: Palgrave MacMillan.

Grenadier, S. (1995) The persistence of real estate cycles. *Journal of Real Estate Finance and Economics*, **10**(2), 95–119.

Hedberg, T. (2020) *The environmental impact of overpopulation.* London: Routledge.

Irwin, T. (2003) *Public money for private infrastructure.* Washington DC: World Bank.

McNellis, J. (2016) *Making it in real estate.* Washington DC: Urban Land Institute.

O'Hara, K. (2014) *Earth resources and environmental impacts.* New York: Wiley.

Stein, D. (1992) *Winning community support for land use projects.* Washington DC: Urban Land Institute.

CHAPTER 5

Project preparation

Preparation of documents

As part of project preparation, the grantor prepares the following tender and other documents:

- Request for Qualification;
- Draft public–private partnership (PPP) contract;
- draft permits and approvals;
- Request for Proposal;
- Instructions to bidders; and
- Response Package.

We discuss these documents below.

Request for Qualification

The Request for Qualification (RFQ), also called the Invitation for Expressions of Interest, is part of the grantor's initial screening process. The RFQ stipulates the project goals, scope of works, procurement strategy, broad expected schedule, and shortlisting criteria. It then invites potential bidders to respond.

Firms that respond to the RFQ will often be evaluated on financial capacity, technical expertise, project experience, details of key project team members, list of major plant and equipment (if relevant), record of disputes, and recommendation letters. Often, they will sign a non-disclosure agreement.

Some governments may impose additional criteria such as domestic project experience, inclusive participation, labor rights (Corvaglia, 2020), sustainable procurement (Addis and Talbot, 2001), and the use of local contractors and operators.

During the process, there are likely to be queries from potential bidders. Firms issue formal queries through Request for Information (RFI) forms. The grantor will need to respond and ensure that critical information is shared among all invited parties to ensure fair bidding. Such information should not be shared informally.

Normally, grantors use a "points system" based on the above criteria to shortlist potential bidders. Suppose the criteria are financial resources (F), technical expertise (T), experience (E), and other factors (O). The grantor may subjectively decide that the weights are 20%, 30%, 30%, and 20%, respectively. Note that the shortlisting criteria differs from the tender award criteria discussed in Chapter 6.

An alternative method uses pairwise comparisons among the criteria (Saaty, 1980) by first constructing the pairwise comparison table:

	F	T	E	O
F	1	**4**	3	5
T	¼	1	2	4
E	⅓	½	1	3
O	⅕	¼	⅓	1
Total	1.78	5.75	6.33	13

For example, if we decide that F is 4 times more important than T, then the (F, T) cell is given the value 4, as shown in boldface font. The (T, F) cell is then given the inverse of 4, which is ¼. By proceeding this way for other pairwise comparisons, the table is completed as shown.

The next step is to normalize the cell values by dividing each element by its column sum to obtain

	F	T	E	O	Weightage
F	0.56	0.70	0.47	0.38	0.53
T	0.14	0.17	0.32	0.31	0.23
E	0.19	0.09	0.16	0.23	0.17
O	0.11	0.04	0.05	0.08	0.07

For example, for the first cell, $1/1.78 = 0.56$. The weightage in the final column is the row average. We may round off the weights as 50%, 25%, 15%, and 10%, respectively.

Draft public–private partnership contract

The draft PPP contract will depend on whether the grantor provides a concession (e.g. for a toll road), is an off-taker (e.g. purchase of desalinated water), or supplies the input (e.g. supply of solid waste to the waste-to-energy plant). It may contain the following:

- obligations of the grantor;
- obligations of the special purpose vehicle (SPV);
- fee mechanism;
- safety and environmental requirements;
- *force majeure* arrangements;
- restrictions on transfer of shares by SPV;
- applicable legal and tax rules;
- step-in rights;
- dispute resolution; and
- possible assignment of benefits to the lender.

The obligations of the grantor may include

- grant of exclusive concession to the SPV for the stipulated period;
- site acquisition;
- facilitation, provision or approval of permits and approvals;
- environmental assessments;
- enabling legislation;
- building of connecting and ancillary services;
- guarantees and tax concessions;
- fuel supply and purchase of output;
- possible compensation to SPV for certain risks (e.g. change in law);
- the right of the SPV to terminate the project with compensation on default of the grantor; and
- supervision and monitoring of progress.

The obligations of the SPV may include

- taking over of the site from the grantor;
- financing, design, and building of the facility to specification by a certain date;
- meeting output quality and volume;
- entering into off-take purchase contract or input contract;
- meeting maintenance standards with penalties for poor performance;
- allowing for early termination with compensation or upon default;
- training of the grantor's workforce before the transfer of asset; and
- ensuring satisfactory asset condition at the end of the contract period.

Normally, the grantor will seek legal opinion in drafting the PPP contract. Every project is different, and the contract is likely to contain special provisions or clauses.

Draft permits and approvals

The grantor prepares draft permits and approvals that are often distributed as part of the tender documents. These documents contain matters pertaining to town planning, building, traffic, safety, health, hazardous waste, pollution, and other environmental issues. A town planning permission concerns land use, such as the type of development, plot density, setbacks, and car parking requirements. A building permit is required before construction of the facilities can begin.

Request for Proposal

The Request for Proposal (RFP), or Invitation to Tender, provides information on the background of the grantor, project goals and objectives, project requirements, procurement strategy (i.e. type of PPP contract), draft contract, copies of permits and approvals, technical information, bid instructions, evaluation criteria, and instructions on how the grantor will handle queries from potential bidders.

For PPP projects, the awarding of the contract is based on the lowest bid if the grantor has a standard design and is looking for the lowest cost.

In many cases, the criteria are design (technical) and price. The criteria for *shortlisting* bidders (financial resources, technical expertise, experience, and other factors) are usually not used for evaluating the tender, which often focuses on design and price. Consider the bids below:

Bidder	Design	Price ($m)
A	70	150
B	80	140
C	50	130
D	55	120

One approach is to consider rejecting bids with design scores of less than 60 out of a possible 100, which eliminates bidders C and D. Then the grantor selects the bidder with the lowest price, i.e. B.

Another possibility is to eliminate bidders C and D as before. The price for the remaining bidders is then normalized by setting the lowest price to 100. The price score (PS) of bidder A is given by

$$PS = 100 - \frac{150 - 140}{140}(100) = 100 - 7.1 = 92.9.$$

Bidder	Design	Price ($m)	Price score	Weighted score
A	70	150	92.9	81.35
B	80	140	100	90.00
C	50	130		
D	55	120		

The weighted score for bidder A, assuming equal weight for design and price, is $0.5(70) + 0.5(92.9) = 81.35$. For B, the weighted score is $0.5(80) + 0.5(100) = 90$. Hence, B is the winner bidder.

If desired, the grantor may change the weightage for design and price. The winning bid should be within 15% of the estimated project cost. Otherwise, the grantor will consider the next best bid, and so on.

Instructions to bidders

These instructions assist bidders on how to respond to the RFP. They include items such as submission requirements, cost of submission, permissible language to use (especially for international tenders), closing date, withdrawal and modifications of offers, clarification procedure, tender deposit, validity of offer, no obligation for the grantor to award the contract, confidentiality of information, conflict of interest, corrupt practices (Guraka, 2016), handling of bid errors, and protests.

Corrupt practices concern issues such as a lack of transparency, procedural irregularities, interference, unauthorized deviations from approved processes, collusion among bidders, kick-backs, and bribery.

Response Package

The Response Package is a formal submission format for the tender, e.g.:

- bidder's particulars;
- price proposal;
- design or technical proposal;
- declaration of non-collusion;
- tender deposit;
- in-principle financing support; and
- proposed contractor, key subcontractors, operator, and suppliers.

At this stage, the grantor is intending to sign a PPP contract with a winning bidder for a PPP project. The bidder may be a consortium comprising the contractor, operator, and other shareholders. Importantly, note that the PPP contract is not a contract to appoint a contractor. As we shall see later, the winning bidder will set up an SPV and subsequently appoint the contractor.

Land acquisition

If the project requires a site that the grantor does not own, the final step in project preparation is site acquisition.

Even if there is a Land Acquisition Act to acquire land for public purposes, this can be a tedious and long-drawn affair over its alleged public purpose, trust, transparency, efficiency of the process, clarity of land ownership, and fair compensation (Chakravorty, 2013).

Within the public sector, it may also not be easy to acquire land from other ministries and agencies because of turf politics (Pressman and Wildavsky, 1979). Hence, land acquisition has to start early so as not to delay the project.

References

Addis, B. and Talbot, R. (2001) *Sustainable construction procurement*. London: CIRIA.

Chakravorty, S. (2013) *The price of land: Acquisition, conflict, consequence*. London: Oxford University Press.

Corvaglia, M. (2020) *Public procurement and labor rights*. Portland, Oregon: Hart Publishing.

Guraka, E. (2016) *Politics of favoritism in public procurement in Turkey*. London: Palgrave Macmillan.

Pressman, J. and Wildavsky, A. (1979) *Implementation*. Berkeley: University of California Press.

Saaty, T. (1980) *The analytic hierarchy process*. New York: McGraw-Hill.

CHAPTER 6

Tender

Pre-qualifying bidders

After project preparation has been completed, the grantor will invite potential bidders to express their interests and pre-qualify for the project by responding to the Request for Qualification (RFQ).

As discussed in the previous chapter, it is common for grantors to use a points system to shortlist about four to eight bidders in a selective tender based on experience, financial resources, and so on. Preparing for a public–private partnership (PPP) tender is costly and time-consuming. Beyond eight bidders, the probability of winning is too small to encourage active participation. It may even draw risky bidders.

Request for Proposal

After shortlisting, potential bidders will be asked to respond to the Request for Proposal (RFP). For a PPP project, this stage may last up to a year or more, depending on the complexity of the project.

Bidders are likely to form consortiums to share risks, expertise, and resources. Often, they will need to provide separate design and price proposals. At this stage, the design is conceptual or preliminary.

Finally, bidders will develop the business case to secure in-principle financing approval. The next chapter will cover how tenderers prepare their bids by performing these tasks.

Responding to queries

During the bidding period, the grantor will respond to queries and clarifications from bidders, sometimes called the *market feedback period*, and not to be confused with market sounding during the feasibility study.

Bidders will provide their feedback and queries, as well as those from potential lenders. For example, lenders may express certain concerns that may affect bankability or make suggestions to improve it. They may also express their concerns with certain contract clauses.

The grantor will hold a pre-bid meeting or briefing for potential bidders near or within the site, followed by a site tour.

Issue of final tender

The grantor will incorporate the relevant feedback from shortlisted bidders and other stakeholders before issuing the final tender.

Alternatively, there is no feedback period or issue of the final tender. The grantor issues the tender documents to shortlisted bidders for their responses. For complex infrastructure projects, it is advisable to have a feedback period and then issue the final tender.

Tender deposit

The tender deposit or bid bond ensures that bidders will honor their bids. It is to cover the loss if the winning bidder decides not to proceed with the project, and the grantor has to award the contract to another bidder or re-tender. A winning bidder may walk away from the project for various reasons, such as the withdrawal of a co-sponsor, the discovery of major bid errors, or the perceived inability to secure higher debt financing.

The bidder purchases the bid bond from the surety or bonding company. The penal sum is the amount the bond will cover, which is usually 5–10% of the bid amount, but it can be as high as 20%. It is the estimated price difference between the winning bid and the second bid. The surety conducts extensive checks on the bidder before agreeing to provide the bond.

The bond premium is about 1–5% of the penal sum, depending on the size of the project and reputation of the bidder (Russell, 1999).

Reference

Russell, J. (1999) *Surety bonds for construction contracts*. Virginia: ASCE.

CHAPTER 7

Sponsor's bid preparation

Decision to tender

In the previous chapter, we discussed how the grantor shortlists potential bidders and issues the Request for Proposal (RFP). Bidders will respond to the RFP by first providing inputs during the market feedback period and attending the pre-bid meeting and site visit. Thereafter, the grantor will issue the final tender after considering the feedback.

A sponsor's decision to tender depends on

- expected profitability;
- bankability;
- the size, uniqueness, and prestige of the project;
- likelihood of finding co-sponsors, local partners, and other investors to share the risks as well as financial and other resources;
- current workload and project schedule;
- time required for bidding;
- cost of bidding;
- transparency of the tender process;
- number of bidders;
- reputation of the grantor, for example, as an off-taker;
- commercial risks; and
- non-commercial risks relating to political, regulatory, social, and legal issues.

Sponsors sign a Development Agreement among themselves on how to proceed. If they subsequently secure the project contract, the agreement forms the basis to develop the Shareholders' Agreement.

There may be one or more sponsors, and the latter is more common because of the complexity of the project and the huge investment costs. We will use the terms "sponsor" and "sponsors" interchangeably.

Development Agreement

The agreement will spell out the basic rules of collaboration, such as the formation of a study team, how to develop the design and price proposals as responses to the grantor's RFP, funding of pre-development activities, and the subsequent shareholding structure and management of the special purpose vehicle (SPV) or project company. The SPV is established after the contract award and financial close.

Project team

Sponsors may have their project and design team or hire external consultants based on expertise, experience, proposed fee, and other criteria such as knowledge of local conditions. At this stage, the grantor has not awarded the contract, and the sponsor will want to minimize the considerable bid preparation costs, which is about 0.5–1.5% of the bid (KPMG, 2010).

The project team for large infrastructure projects consists of senior members with extensive expertise and experience. These are the hard skills discussed in Chapter 2. There are clear roles and responsibilities in the project governance structure. The team members regulate their relations by contract and develop soft skills (see Chapter 2) to build trust and get along. It also develops internal and external coalitions to get things done.

Political, regulatory, and legal assessments

In Chapters 3 and 4, we discussed how the grantor assesses the political, regulatory, and legal risks. The SPV carries out a similar assessment from a different perspective. These risks to the SPV include

- interference in the project;
- revocation of contract;

- confiscation of assets;
- non-payment by the grantor as an off-taker or buyer of the project output;
- new regulatory requirements;
- weak enforcement of contracts and dispute resolution mechanisms;
- disruption in supplies;
- forced participation by state-owned enterprises;
- non-approval of permits and licenses;
- change in taxation;
- trade curbs;
- corruption;
- deliberate delays;
- renegotiation of contract on less favorable terms; and
- currency controls.

There are ways to mitigate these non-commercial risks, such as by signing a government support agreement on the above, borrowing from international multilateral banks to discourage governments from predatory behavior, and the use of appropriate contract clauses such as international arbitration.

Design proposal

The design proposal is guided by the grantor's aspirational and functional objectives. This is supplemented by the project team's inputs to try and win the bid. The aspirational aspects may include new technologies, methods, benchmarks, designs, and uses for the site. The grantor is likely to provide the development concept in the tender brief without over-constraining how bidders will respond.

Depending on the type of project, the functional items include the scope and size of the project, types of spaces, integration, flexibility, energy efficiency, sustainability, and so on. These items are found in the tender specifications. For example, a high-speed rail (HSR) system consists of the following components:

- infrastructure;
- rolling stock (trains); and
- operation and maintenance (O & M).

The infrastructure includes the route layout, civil works, stations, depots, track geometry, power supply, and signaling system. The scope of O & M includes timetabling, scheduling, ticketing, maintenance, and repairs. The grantor is likely to award separate contracts to private contractors to build the infrastructure because of the scale and complexity of the project. Similarly, stations, depots, power supply, and the signaling system can be unbundled into separate contracts. There may be a single or separate O & M contracts.

Route planning requires information on expected ridership based on locals and tourists. Once the tentative stations and routes have been selected based on demand studies, the grantor will often appoint external consultants to

- conduct preliminary engineering studies;
- provide an environmental impact assessment (EIA);
- integrate the proposed HSR with existing rail lines and infrastructure; and
- study the topography and geology.

The grantor uses the above information to evaluate the possible routes by considering the benefits and costs as well as other criteria. Thereafter, geotechnical investigations can begin to finalize the route.

If there are separate O & M contracts, the SPV, as the operator, needs to submit proposed designs for the timetable, schedule, ticketing, and so on. At the tender stage, the design is conceptual or schematic. It breaks the system into components and their respective narratives. As the design progresses, there will be progressive cost estimates to develop the price proposal.

Price proposal

The sponsor's price proposal concerns profitability; that is, the project is commercially viable and hence able to attract financing; that is, it is bankable. The profitability of a project depends on many factors, as discussed below.

Revenue

Sponsors can mitigate demand risks by

- conducting a proper market study;
- using "take or pay" purchase contracts;
- using financial options;
- seeking creditworthy buyers; and
- ensuring that the product meets quality standards.

We discuss the first three methods below.

Market study

A market study is used to estimate revenue, that is, output demand and price. Demand depends on price as well as shifters such as demographics, income, price and availability of substitutes, and policy variables such as taxation and exchange rates. These variables shift the demand curve. Recall that the demand curve shows the relationship between quantity demanded (X-axis) and price (Y-axis).

The price depends on demand and supply. On the supply side, competitors may react through price and non-price competition. For non-durable goods, the intersection of the demand and supply curves determines the equilibrium price. For durable goods, the *stock-flow model* is used for short-term forecasting. Longer-term projections are based on growth rates of demography, income, and other variables.

Stock-flow model

As an example, we will consider a stock-flow model of the private housing market. A housing unit (or simply a house) is a stock concept. It is a durable good that lasts a long time (e.g. 99 years) and provides a flow of continuous housing services. Similarly, a hammer provides a flow of hammering services, and a refrigerator provides a flow of refrigeration services. Hence, stocks are physical housing assets and should not be confused with the stock or equity market for company shares. The term "stock" is also used for business inventory, as in "no more stock" or "stock-taking."

The market for housing assets is where houses are traded; for example, a house may sell for $500,000. The market for housing services is the rental market. For example, the same house may rent for $20,000 a year, giving a gross yield of $20,000/$500,000 = 4%. The net yield is obtained by subtracting outgoings from the gross rent before dividing it by the house price.

There are many types of houses, such as public flats, private apartments, mansions, and bungalows. We simplify and call them "houses" or standardized "housing units." In aggregate, a city of 4,000,000 residents has a housing stock of about 4,000,000/4 = 1,000,000 housing units, assuming an average household size of 4. Each year, the city may build 20,000 housing units. This is the flow of housing units that is added to the stock each year. Observe that the annual flow of housing units is 20,000/1,000,000 = 2% of the housing stock.

The demand for private housing units (D_t) depends on price and other non-price demand "shifters," such as the rate of household formation, household income, credit conditions (e.g. loan to value ratio), mortgage interest rate, prices of substitutes (e.g. public or rental housing), exchange rates (for foreign buyers), and policy variables such as taxes, stamp duty, and other housing regulations. The demand curve is a function of price only, and changes in non-price variables shift the demand curve; that is,

$$D_t = f(P_t, \mathbf{x}_{t-s})$$

where P_t is the house price at time t, $f(.)$ is usually a linear or log-linear function, and $\mathbf{x}_{t-s}$ is a vector of the shifters. A linear model has the form

$$y = \alpha + \beta x_1 + \phi x_2$$

and a log-linear model is given by

$$log(y) = log(\alpha) + \beta log(x_1) + \phi log(x_2).$$

Generally, $\mathbf{x}_{t-s}$ contains current and lagged variables because the sale and purchase of a house is a long process. For example, since buyers arrange their financing about one quarter prior to the actual sale, the mortgage interest rate (r) should be lagged by one period, that is, r_{t-1}.

There may be another quarter of lag between a sale and the lodgment of the caveat. Lags are not required if we use annual time series data.

The supply of housing units (S_t) in any period consists of the previous stock (S_{t-1}) and newly completed units (C_t) so that

$$S_t = (1 - \lambda)S_{t-1} + C_t.$$

Here, λ is the annual rate of physical depreciation, which is about 2% or 0.02 for residential buildings. This assumes that a typical house is fully depreciated in 50 years. In addition, the supply of housing units includes additions (renovations) and losses through demolitions, floods, and fire. The standard assumption is that these effects cancel out and are, hence, left out of the equation.

The supply of new units (C_t) depends largely on the actions of housing developers, that is,

$$C_t = g(P_{t-k}, \mathbf{y}_{t-k})$$

where $g(.)$ is the linear or log-linear supply function, P is the house price as before, and $\mathbf{y}$ is a vector of supply shifters, such as land cost, construction cost, interest rates, and policy variables that affect housing supply. These variables include the developer's stamp duty, taxes, credit rationing, and land use constraints. Because housing construction takes time, developers make their decision to build about one to three years earlier, which explains the presence of lag k. For example, if there is a lag of one year, $k = 1$. Developers build on forward-looking expectations that houses are sellable in future. This is called speculative building, and is risky.

To solve the above system of equations, we need to specify the market clearing condition. Disequilibrium models are more complicated (Fair, 1972), and many housing models assume market equilibrium so that

$$D_t = S_t.$$

We solve for the house price by substituting the previous three questions into the equilibrium condition, giving

$$P_t = h(P_{t-k}, \mathbf{x}_{t-s}, \mathbf{y}_{t-k}).$$

Usually, $h(.)$ is assumed to be linear or log-linear so that it can be estimated using linear regression (Tan, 2022). The estimated model may then be used to study and forecast house price movements. If desired, it is possible to specify a disequilibrium model by allowing price changes to adjust depending on the lagged vacancy rate (V_{t-1}) and prices plus a random term (ε), that is,

$$\Delta P_t = \alpha + \beta V_{t-1} + \varphi \Delta P_{t-1} + \varepsilon_t$$

where α is the intercept, and β, φ are parameters. There are many variants of such disequilibrium adjustments.

The model requires a lot of data because of the large number of variables over many periods. Some of the data may not be available, up-to-date, or accurate. To estimate these equations, the time series must be stationary; otherwise, first differencing is required for each non-stationary variable.

Purchase contract

Sponsors may also mitigate demand risk by using take-or-pay purchase contracts specifying minimum quantities and prices; that is, the purchaser must pay a penalty (e.g. 80% of the contract price) even if he does not take delivery of the output.

Such contracts are common in the energy sector because of the high initial investment. Without such a contract to guarantee a minimum volume of sales, the investment may be too risky for the seller.

However, outside the energy sector, take-or-pay contracts may not be enforceable. It may be interpreted as a liquidated damages clause, which is deemed to be fairer because the penalty depends on the actual damage rather than on a punitive sum.

Call option

Another way to mitigate demand risk is to use call and put options. A call option allows the holder the right to *buy* an asset in the future at a specified price at the expiry or exercise date. For example, a developer may wish to purchase a piece of land for development for $100,000 (the exercise or strike price). The landowner sells the developer a call option for $2,000 (the premium or option fee) to keep the land off the market for three months for the latter to conduct a feasibility study and arrange for financing. On the expiry date, the developer may exercise the option and pay the remaining $98,000 or forgo the $2,000 if the project is not feasible.

There are variations from this basic call option, such as the ability of exercise at any time (an American option) rather than only at expiry, which is a European option. Option pricing refers to methods of ascertaining the premium, which depends on expectations of the value of the asset at expiry, the duration of the option period, the exercise price, and the time value of money.

As illustrated above, call options are used in projects to purchase inputs, such as land, fuel, and raw materials.

Put option

A put option is an option to *sell* an asset or the project output in the future at a specific price. For example, a landowner is migrating and wishes to sell the land in three months for $110,000. He pays an investor $3,000 for a put option. If the landowner decides to sell the land in three months, the investor will pay the remaining $107,000. However, if land prices spike, the landowner will not exercise the option and lose $3,000. He will sell the land in the open market at a higher price. However, if land prices slide, the landowner will exercise the option.

The investor is thinking that land prices will go up if he enters into a put contract. The landowner believes that land prices will fall and wishes to lock in the price at $110,000.

Option pricing

Option pricing refers to the determination of the premium given the current price, strike price, time to expiry, and volatility. The *Black–Scholes model* (see (McDonald, 2014)) is briefly discussed below for a call option on the price of a company share or stock.

The *volatility* of a stock is the standard deviation (σ) of the rate of return on stock prices. For example, if S represents annual stock prices, the rate of return for any given year is

$$r_t = \frac{S_t - S_{t-1}}{S_{t-1}}.$$

If there are n annual rates of return $r_1, \ldots, r_n$, then

$$\sigma^2 = \frac{1}{n-1} \sum_{i=1}^{n} (r_i - \bar{r})^2.$$

where $\bar{r}$ is the mean rate of return. Taking the square root of the variance gives the volatility σ.

The Black–Scholes model assumes S moves in continuous time (t) according to the *Geometric Brownian Movement* (GBM), that is,

$$dS = \mu S dt + \sigma S dz. \tag{7.1}$$

Here,

$$dz = \varepsilon \sqrt{(dt)}, \quad \varepsilon \sim N(0, 1).$$

The differential dz follows the *Wiener process* or Standard Brownian Movement (SBM). In discrete time, it is called a *random walk* and is given by

$$z_t = z_{t-1} + \varepsilon_t.$$

More generally, we can write Equation (7.1) as

$$dS = a dt + b dz. \tag{7.2}$$

Here, a and b are functions of S and t. For example, comparing Equations (7.1) and (7.2), $a = \mu S$ and $b = \sigma S$. Note that S and dz are functions of t, and the time subscripts are omitted to avoid clutter. Before deriving the Black–Scholes model, we need Itô's lemma.

Itô's lemma

If $f = f(S, t)$ and S fluctuates according to Equation (7.2), then

$$df = \left(a \frac{\partial f}{\partial S} + \frac{\partial f}{\partial t} + \frac{1}{2} b^2 \frac{\partial^2 f}{\partial S^2} \right) dt + b \frac{\partial f}{\partial S} dz.$$

The proof is based on Taylor's expansion of $f = f(S, t)$, that is,

$$df = \frac{\partial f}{\partial S} dS + \frac{\partial f}{\partial t} dt + \frac{1}{2} \frac{\partial^2 f}{\partial S^2} (dS)^2.$$

Substituting for dS from Equation (7.2),

$$df = \frac{\partial f}{\partial S}(adt + bdz) + \frac{\partial f}{\partial t}dt + \frac{1}{2}\frac{\partial^2 f}{\partial S^2}(adt + bdz)^2.$$

By collecting the terms and simplifying, we have

$$df = \left(\frac{\partial f}{\partial S}a + \frac{\partial f}{\partial t} + \frac{1}{2}\frac{\partial^2 f}{\partial S^2}b^2\right)dt + \frac{\partial f}{\partial S}bdz.$$

The above expression uses the following results:

- $dz = \varepsilon\sqrt{(dt)}$, so $(dz)^2 = \varepsilon^2 dt = dt$ because $E(\varepsilon^2) = 1$; that is, the expected value of a chi-square variable equals its degree of freedom; and
- $(dt)^2 = dtdz = 0$ for small increments dt and dz.

This completes the proof.

We are now ready to derive the Black–Scholes model. Let $C(S, t)$ be the value of a call option on S. Then, using Equation (7.1) and Itô's lemma,

$$dC = \left(\mu S\frac{\partial C}{\partial S} + \frac{\partial C}{\partial t} + \frac{1}{2}\sigma^2 S^2\frac{\partial^2 C}{\partial S^2}\right)dt + \sigma S\frac{\partial C}{\partial S}dz. \tag{7.3}$$

If we hold a portfolio comprising x units of bonds and y units of stocks, its value is

$$P = xB + yS. \tag{7.4}$$

Hence,

$$dP = xdB + ydS.$$

All variables in the above portfolio equation are functions of time. If we invest in a bond at a constant risk-free interest rate r, the instantaneous return is

$$dB = rBdt.$$

To understand this expression, rewrite it as

$$\frac{dB}{B} = rdt.$$

Integrating both sides gives

$$B = B_0 e^{rt}$$

where B_0 is the value of the bond at time $t = 0$, and e is the base of a natural logarithm. This expression shows that we are assuming that the value of the bond grows exponentially at rate r over time.

Substituting for dB in the portfolio equation, we get

$$dP = xrBdt + y(\mu Sdt + \sigma Sdz) = (rxB + y\mu S)dt + y\sigma Sdz. \qquad (7.5)$$

If the value of the call option is assumed to be equal to the replicating portfolio, $dC = dP$. Hence, equating Equations (7.3) and (7.5) gives

$$y = \frac{\partial C}{\partial S}$$

$$x = \frac{\frac{\partial C}{\partial t} + \frac{1}{2}\sigma^2 S^2 \frac{\partial^2 C}{\partial S^2}}{rB}.$$

Substituting x and y into Equation (7.4) and setting $P = C$, we obtain the Black–Scholes partial differential equation (PDE)

$$rs\frac{\partial C}{\partial S} + \frac{\partial C}{\partial t} + \frac{1}{2}\sigma^2 S^2 \frac{\partial^2 C}{\partial S^2} = rC. \qquad (7.6)$$

The solution to this PDE is

$$C = SN(d_1) - Ke^{-rT}N(d_2) \qquad (7.7)$$

where

$$d_1 = \frac{\log\frac{S}{K} + \left(r + \frac{1}{2}\sigma^2\right)T}{\sigma\sqrt{T}}$$

$$d_2 = d_1 - \sigma\sqrt{T}.$$

Here, $N(z)$ is the shaded area under the standard normal curve from negative infinity to z. For example, $N(0) = 0.5$, or half the area under the standard normal curve. As before, r is the risk-free interest rate, T is the time to expiry in years, σ is the volatility of the price series S, and $log(.)$ stands for the natural logarithm function.

In summary, the steps are as follows:

- compute d_1 and d_2;
- find $N(d_1)$ and $N(d_2)$ by looking up the standard normal distribution table; and
- compute C using Equation (7.7).

Example: Pricing of a call option

The current land price (S) is $100 per m². If the strike price K is $105 per m² with a $T = 2$ year expiry, price the call if the risk-free interest rate (r) is 0.04 and the volatility (σ) is 0.2.
    Solution:

$$d_1 = \frac{\log \frac{100}{105} + \left(0.04 + \frac{1}{2}(0.2)^2\right)2}{0.2\sqrt{2}} = 0.251$$

$$d_2 = 0.251 - 0.2\sqrt{2} = -0.032.$$

From the standard normal statistical table, $N(0.251) = 0.599$, and $N(-0.032) = 0.488$. Hence,

$$C = 100(0.599) - 105e^{-0.04(2)}(0.488) = \$12.60.$$

Note this is a European option; that is, it is exercised only at expiry. For an American option that can be exercised at any time, valuing the call is more complicated (Joshi, 2010).

Procurement strategy

Recall that the public agency uses the public–private partnership (PPP) procurement strategy to contract with the sponsor's SPV. A sponsor's procurement strategy is different and consists of three elements, namely,

- ways of sourcing for contractors;
- construction delivery methods; and
- payment methods.

Contractors or bidders may be selected through open, selective, or limited tender. For infrastructure projects, open tenders are impractical, considering the high cost of project preparation and complexity of the project. Hence, a common strategy is to use a selective tender by pre-qualifying about six potential bidders. In a limited tender, there are only one or two bidders, such as in sensitive military projects.

There are many construction delivery and payment methods, and these two elements are often discussed together as part of a construction contract. The types of construction contracts are discussed below.

Lump-sum contract

The lump-sum contract is also known as the fixed price, hard-bid, Design-Bid-Build (DBB), or traditional contract. The sponsor (owner) has a site and engages a project team to determine the scope, requirements, schedule, budget, and design. If the owner is a developer, it is cheaper to have a core in-house project team. The composition of the team includes the project manager, designers (engineers and architects), contract manager, and other consultants appointed for cost estimation, financial, insurance, legal, taxation, environmental, energy simulation, and other purposes depending on the requirements of the project.

The design proceeds in three phases: conceptual or schematic design, design development, and the development of construction documents. The end products are the tender documents comprising the specifications, construction contract, detailed drawings, and other documents.

Contractors tender for the project based on the detailed design. The award is based on the lowest price or a points system. This price is fixed, subject to variation or change orders.

During construction, the consultants provide independent project control over the contractor on behalf of the owner. However, the owner will have to handle disputes between the project team and contractor.

The DBB contract provides price certainty (subject to variation orders), which is preferred by lenders. However, it is more suitable for projects where the design or scope is more or less fixed, such as small projects and standard buildings, e.g., schools, public housing, and factories.

Design-Build contract

In a Design-Build (DB) contract, the owner awards a single contract for design and construction rather than separate contracts in the lump sum contract. It is called the Engineering, Procurement, and Construction (EPC) contract in the process industry. The design-build team may be designer-led or contractor-led if it is a joint venture between different entities. Alternatively, it is a firm that specializes in DB projects. Such a firm will have separate in-house design and construction teams.

The DB contract is more suitable if the design is more complex and changes are expected. It provides a single point of responsibility, and designers no longer exercise independent project control over the contractor on behalf of the owner.

Unlike a lump sum contract, there is no design for the DB firms to bid. The owner may

- select a DB firm and negotiate the price before design commences, or
- provide basic specifications and invite several firms to submit design and price proposals for consideration.

Like the public agency, the sponsor may use a composite score to identify the winning bidder.

There are variations in DB contracts. The owner may engage the designers to control the conceptual design before *novating* (transferring) the design contract to the contractor. The contractor then leads the design development process.

The BD contract is also an example of the *Early Contractor Involvement* (ECI), where the contractor is brought in early to provide input to facilitate the integration of design and construction.

In the *Integrated Project Delivery* (IPD) approach, the owner and project team use a collaborative multi-party contract to design and build a facility to target cost. The owner negotiates and selects the team based on quality, experience, financials, and other criteria. The parties share the risks, and payment depends on performance rather than price competition. However, the parties may not work well together, and IPD may not attract financing.

For large infrastructure projects, lenders prefer fixed price contracts and tested clauses. The owner carries the design risks in DB contracts. Once the design is agreed, it can be expensive for the owner to change the design. Further, the DB firm may offer a design that is easy to build and sacrifice the aesthetics.

Construction Management contract

In Construction Management (CM) or agency CM, the owner engages designers to design the project and a construction manager to implement it for a fee. The owner signs different contracts with the designers and specialist subcontractors, and hence takes the risks of project delays and cost over-runs.

A variation of this project delivery method is CM at-risk. Instead of a fee, the construction manager signs the subcontractor contracts and assumes the construction risks. Hence, CM at-risk is similar to DBB, but the construction manager is brought in early during the design phase.

Measurement contracts

Measurement contracts, also called re-measurement or unit price contracts, are used if the design is reasonably detailed but quantities are uncertain, such as in civil works. Instead of a fixed lump sum, contractors bid by furnishing unit rates for various items of work. Payment is based on the actual quantities used in the project.

A variation of the unit price contract is the percentage rate contract, where the owner provides the unit rates (p_i) and estimated quantities (Q_i) and uses them to estimate the construction cost (C); that is,

$$C = \sum_i p_i Q_i.$$

Bidders then bid by providing a percentage above or below C, such as $C + 5\%$. The additional 5% is then distributed across the unit rates. As before, payment is based on actual quantities.

Cost estimates

Sponsors (owners) tend to use per unit basis, such as $ per hospital bed or $ per km of road, to estimate project costs from the preliminary design. Unlike cost-benefit analysis, where competitive prices are used, sponsors will use actual prices to compute the profitability of the project. For example, if utilities are subsidized, sponsors will use the subsidized price. There is no requirement to compute shadow exchange rates and shadow prices. Taxes and interest expenses are also part of a sponsor's project costs.

For external costs such as air pollution, sponsors will need to comply with regulatory requirements.

In the process industry, cost capacity curves may be used to estimate the cost of a proposed plant (C) with capacity Q using the *parametric method*:

$$C = c \left[\frac{Q}{q} \right]^{\alpha}.$$

Here, c and q are the cost and capacity of an existing comparative plant, and α is a parameter determined from historic data.

If further information is available, a more accurate preliminary cost estimate is possible, such as the following for a road:

- length of road;
- number of lanes;
- type of surface material, e.g. bitumen or concrete;
- method of construction;
- number of bridges and tunnels;
- terrain and types of soils; and
- designed traffic load.

For a building, the *elemental cost approach* may be used if there is sufficient information on the building items or elements (Table 7.1).

Table 7.1   Elemental cost estimate.

Item	Existing building		Proposed building	
	Cost	Cost/m^2	Cost/m^2	Cost
General requirements	200,000	500	510	220,000
Site work				
Concrete				
Masonry				
...				
Electrical				
Total cost	3,500,000	3,000	3,100	3,800,000

Schedule estimate

Sponsors will also estimate the schedule using a bar chart comprising the pre-construction, construction, and post-construction activities. At this stage, the schedule is preliminary and relies primarily on experience in executing similar projects.

The pre-construction activities include a feasibility study, preparing and bidding for the PPP contract, project design, securing early regulatory approvals, developing the construction procurement strategy, and mobilizing resources for construction.

The construction activities consist of achieving major milestones such as the completion of major building components, commissioning of major systems, training of operatives, and occupation. Prior to construction, the appointed contractor will furnish a master construction schedule based on critical path analysis (see Chapter 11).

Finally, the post-construction period includes testing, calibration, rectification of defects, and ramp-up towards full operational capacity.

Quality

Quality has different meanings, such as value for money, fit for purpose, free of defects, conformance to standards, meeting customer needs, or as a process; that is, doing things right the first time.

Sponsors control project quality through a quality plan that includes contractor selection, supplier selection, specifications, quality assurance of processes, quality control, retention sum, and warranties against defects. The measures are incorporated into the construction contract. The retention sum or security deposit is about 10% of the monthly progress payment to the contractor, up to a limit of 5% of the contract sum. Upon the certification of the completion of construction by the sponsor's representative, half the retention sum is released to the contractor. If the contractor does not rectify defects during the defects liability period (DLP) after the completion of construction, the sponsor may use the remaining retention sum to pay for rectification works.

Similarly, the contractor is, by contract, required to furnish a quality plan for sponsor's approval. Part of the plan concerns how the contractor selects and monitors subcontractors to ensure quality.

Construction

The contractor may not perform in terms of time, cost, quality, and safety. Hence, the procurement strategy and choice of contractor are important considerations. In addition, sponsors use incentives and penalties such as early completion bonus, performance bond, warranties, and liquidated damages to motivate the contractor. The performance bond should cover any extension of time and the defects liability period. Its value will need to be adjusted if there are major contract variations.

Sponsors need to put in place a rigorous method of project management and control to avoid scope creep, delays, construction accidents and safety violations, cost escalations, claims, and costly disputes.

Inputs

A project requires inputs such as raw materials, product components, energy, and utilities. Input supplies require site access. A major risk with inputs is fluctuating quantities and prices, and the SPV may or may not be compensated for these movements. For electricity generation projects using oil or gas as fuel source, there is often a "pass-through" provision to pass the risk to consumers. However, if the SPV is building a road, there is often no such clause for sand and other inputs.

The SPV may use forward purchase contracts or futures contracts to hedge against fluctuating input prices. Alternatively, it may purchase a call option to buy a certain quantity of the input at a fixed price in future. Recall that a call option gives the holder the right, but not the obligation, of exercise before or at expiry.

Another hedging strategy is to diversify input sources to ensure prices are competitive and reduce supply disruptions.

Forward contracts

The SPV, as a buyer, may enter into a forward purchase contract with a seller to secure a certain input at fixed prices at a future (exercise) date, such as when the input is required during the operational phase. The SPV may enter into such supply contracts for only a portion of the project input needs (e.g. 60%) so that it can also benefit by buying from the open market if market prices fall below the fixed contract price. If market prices rise on the exercised date, the SPV loses money on 60% of the input, but it can sell the other 40% in the open market to cover the loss. Both parties may settle through delivery of the commodity or, more likely, through cash. The latter is computed as (spot price minus contract price) times the number of units.

To illustrate the principles, suppose it is January now, and a farmer intends to produce 10,000 units of a commodity during harvest at the end of June, the exercise date. The current price is $100 per unit. The farmer enters into a forward contract with a buyer to sell 6,000 units at $110 per unit. If the spot price rises to $120 per unit at the end of June, the revenue for the farmer is $1,140,000, as shown below. The spot price is the prevailing price in the market.

	End-June	Quantity	Contract value
Contract price per unit	$110	6,000	$660,000
Spot price	$120	4,000	$480,000
Total revenue			$1,140,000

However, if the spot price falls to $80 per unit, the farmer's revenue is $980,000, as shown below.

	End-June	Quantity	Contract value
Contract price per unit	$110	6,000	$660,000
Spot price	$80	4,000	$320,000
Total revenue			$980,000

If the farmer does not hedge against price fluctuations, the revenue can fall as low as $80(10,000) = $800,000 or as high as $120(10,000) = $1,200,000.

Similarly, the buyer (SPV) will enter into a forward contract to purchase part of the inputs. The logic is the same; the buyer has the opportunity to purchase the input from the spot market if prices fall below the contract price.

Suppose the spot price is $120 per unit at the end of June. Then, the seller will pay the buyer ($120 − $110)6,000 = $60,000. The buyer will use this amount to purchase the input from the spot market at a higher price of $120 per unit.

By entering into a forward contract, both the buyer and seller have hedged their risks. However, forward contracts have some downsides. The system is unregulated. The seller needs to find a buyer, and vice versa. Further, the seller cannot be assured that the buyer will not default on the payment.

Futures contract

A futures contract is similar to a forward contract except that it is traded on an exchange. Hence, there is no need to physically find a buyer or seller. Further, the exchange ensures that the counter-party will not default on the payment.

To facilitate the trading of futures contracts, the contracts are traded in standardized units, such as 1,000 units. Futures contracts are quoted on a monthly basis, and the exercise date is, for example, the last working day of the month. For example, if it is April now, the bids and offers for a June futures contract may look like this:

Bid ($)	Quantity	Offer ($)	Quantity
105	1	106	2

The contract quantities are in lot sizes, and 1 lot may be 1,000 units. There are offers to sell the futures contract expiring at the end of June for $106 per unit for 2 lots. If an investor thinks that the price will rise, he will purchase 1 lot for $106 per unit. Otherwise, he may queue at $105 per unit and wait for a seller to sell at this price. For this example, assume that the purchase price is $106 per unit.

If the spot price of the commodity at the end of June is $120 per unit, the buyer gains because he has purchased it for $106. However, if the price falls to $80, the buyer of the contract loses money.

Operation and maintenance

Many things can go wrong during the operation and maintenance (O & M) phase, which is the longest period of a project's life cycle. There are two main risks, namely, inefficiency and a failure to perform.

Clearly, having an experienced operator helps, particularly if he is also motivated by his equity commitment in the project. However, having the operator as an equity investor poses a possible conflict of interest. Hence, there is a need for performance incentives and penalties for poor performance.

Generally, sponsors and lenders prefer tried and tested technology. If the project uses a new technology, the SPV must secure a guarantee from the technology supplier. Other equipment may also fail, and a warranty is necessary. It is also possible to enter into a maintenance agreement with the manufacturer.

Finally, facilities managers and operatives require training by experienced facilities managers, equipment suppliers, and technical experts.

Force majeure events

There are three forms of *force majeure* (FM) risks, namely,

- political and social unrest such as wars, strikes, and civil unrest;
- change in law; and
- natural disasters and other events (e.g. an explosion, a pandemic).

In general, the SPV is excused from contractual performance for political FM and a change in law, and *may* be entitled to time extension and compensation by the grantor depending on the terms of the contract. However, it needs to insure against other FM risks.

Sources of funds

Sponsors often finance their projects with equity and debt. Although sponsors provide "equity," they usually provide subordinated or shareholders' loans that are often treated as equity. Instead of receiving dividends, they receive interest payments, which the SPV can book to reduce its tax.

Passive investors may provide equity or purchase convertible debt; that is, it is convertible to equity if the project is successful.

To get the best deal, sponsors invite lenders to bid for the loan package. These lenders form a syndicate among themselves to share the loan and reduce their exposure to risks. These senior bank loans have priority for claims over shareholder loans if the SPV defaults or goes bankrupt.

In some countries, such as the United States (US), another institutional lender may "take out" the construction loan during the operation phase. This mechanism requires the permanent lender to issue a commitment letter to the sponsor and construction lender. There are two reasons for this arrangement. First, the construction lender does not want to lend on a long-term basis to reduce the mismatch between the short-term nature of its savings deposits and long-term lending. Second, the permanent lender, such as an insurer that invests in the long term to match its insurance policies, does not want to handle the construction risk.

During the operation phase, it is possible to refinance the loan because interest rates may have fallen or there is no longer any construction risk. With lower risks, borrowing rates should come down. The refinancing may be through another bank loan or by issuing a bond. The latter is more complicated because the SPV has to deal with many different investors, issuance costs and procedures, greater disclosure, and subsequent difficulties in adjusting financing terms and covenants.

Other sources of funds include

- export credit agencies that provide loans related to the export of their countries' equipment to the project;
- short-term supplier credit;
- passive investors such as infrastructure investment funds, pension funds, and insurance companies;
- government grants as a catalyst to encourage participation by private investors;
- carbon credits;
- multilateral development agencies, such as the World Bank or Asian Development Bank; and
- Islamic finance.

The World Bank's Multilateral Investment Guarantee Agency (MIGA) provides political risk insurance, a partial credit guarantee to cover the sovereign borrower's non-payment of debt in public projects, and a partial risk guarantee to cover non-performance of sovereign contractual obligations.

Islamic finance differs from conventional loans. Financing is based on Islamic laws that prohibit the payment or receipt of interest and lending for certain activities such as gambling and alcohol. The lender provides a loan and, instead of interest payments, he takes a share of the profit. Since it is possible for the borrower not to earn a profit, the lender effectively shares the risks.

Islamic finance is an important source of funds for projects in countries not well served by conventional project finance. A project may also be jointly funded by conventional and Islamic finance.

Cost of funds

Recall from Chapter 3 that the weighted cost of capital (*WACC*) is given by

$$WACC = (1 - \phi)r_e + \phi(1 - t)r_d.$$

Here, ϕ is the fraction of debt, r_e is the cost of equity, r_d is the cost of debt, and t is the corporate tax rate (e.g. 0.25). For example, if a project uses 30% equity and 70% debt, then $\phi = 0.7$. Generally, interest expense on debt is tax deductible, which explains why the cost of debt is lowered by $(1 - t)$.

What determines the value of ϕ? This is the issue of *optimal capital structure*, the mix of debt and equity that minimizes *WACC*. By minimizing *WACC*, the firm maximizes its market value because *WACC* is used to discount future cash flows. There is an outdated irrelevance theory that postulates that the riskiness of a firm depends on the type of business and not on capital structure (Durand, 1952).

From the sponsor's point of view, it is cheaper to take more debt because of the tax treatment. However, taking more debt makes the firm riskier, which increases its cost of equity. Hence, the firm should increase its debt up to a limit.

This raises the issue of agency conflict or the separation of ownership and control of the corporation (Berle and Means, 1932). Shareholders (sponsors) employ managers to manage the firm. If the compensation of managers includes equity, they tend to be conservative and ignore higher-risk projects that have positive NPVs.

More generally, shareholders, lenders, and managers may have different objectives when it comes to the optimal capital structure. Lenders want to see strong sponsor commitment to the project, which means considerable equity commitment by sponsors. There should also be adequate debt service cover to service the loan; that is, net earnings must adequately cover periodic debt repayment. In practice, ϕ is about 0.7 for infrastructure projects.

The cost of a bank loan (r_d) is the best rate quoted by competing syndicates for the project loan package. If the SPV issues a bond during the operation phase, the cost of the bond depends on the coupon payment. For example, the cost of a $1,000 bond that pays $50 a year is 5%. It needs to be adjusted for issuance cost and corporate tax.

The cost of equity (r_e) may be estimated using Gordon's formula. Suppose a company's stock with similar risks to the SPV earns a dividend D_1 at the end of the first year, and this dividend grows at the rate of $g\%$ per year. The current value (V) of the share is the present value of future dividends, that is,

$$V = \frac{D_1}{1+r_e} + \frac{D_1(1+g)}{(1+r_e)^2} + \cdots = \frac{D_1}{r_e-g}.$$

Hence,

$$r_e = g + \frac{D_1}{V}.$$

This shows that the return to equity may be estimated as the sum of the dividend growth rate and the first-year dividend yield. For example, if the company's share currently trading at $1.20 provides a dividend of 6c every year, its cost of equity is $r_e = 0 + (6/120) = 0.05$ or 5%.

An alternative approach to estimating r_e is to use the Capital Asset Pricing Model (CAPM), that is,

$$r_e = r_f + \beta(r_m - r_f).$$

Here, r_e is the expected cost of equity, r_f is the risk-free interest rate, β is the "beta" for the specific company, and r_m is the expected market rate of return. The value of r_m is estimated from the stock market index, which is based on the share prices of a basket of selected companies from different economic sectors. For example, the Straits Times Index tracks the performance of 30 large companies listed on the Singapore Stock Exchange. Some texts use $E[r_e]$ and $E[r_m]$ to make it explicit that they refer to expectations. The expected value of a variable X, written as $E[X]$, is the population mean. The risk-free rate is approximated by the rate on long-term government bonds (e.g. 2.5%). The term $(r_m - r_f)$ represents the *market risk premium*. Hence, $\beta(r_m - r_f)$ represents the risk premium for the company's share. If $\beta = 1$, the company's share is as risky as that of the market. If $\beta < 1$, the company's share is less risky; that is, it is a defensive stock. If $\beta > 1$, the company's share is riskier than that of the market index. For example, if $\beta = 1.2$, then

$$r_e = r_f + \beta(r_m - r_f) = 2.5\% + 1.2(5\% - 2.5\%) = 5.5\%.$$

To find the value of β for a company, we rewrite the CAPM model as

$$(r_e - r_f) = \alpha + \beta(r_m - r_f) + \varepsilon.$$

This is the same as the simple regression model

$$y = \alpha + \beta x + \varepsilon.$$

Here, α is the intercept, and ε is the error term. Stockbrokers use monthly data to estimate and update the betas of many listed companies.

Amortizing loan repayment

For a L loan (the principal) at an annual interest rate i, the *annual* constant repayment amount to amortize or fully repay the loan in n years is

$$Z = \frac{Li}{H}$$

where $H = 1 - (1 + i)^{-n}$ and n is the amortization period. The *monthly* repayment is

$$Z_m = \frac{L\left(\frac{i}{12}\right)}{G}$$

where

$$G = 1 - \left(1 + \frac{i}{12}\right)^{-12n}.$$

Observe that $Z \neq 12Z_m$ or $Z/Z_m \neq 12$; that is, 12 monthly repayments do not equal one annual repayment. The correct relation is

$$\frac{Z}{Z_m} = \frac{Li/H}{L\left(\frac{i}{12}\right)/G} = \frac{12G}{H}.$$

Clearly, the ratio is not 12 unless $G = H$, which is unlikely in practice.

Consider a loan with $L = \$1,000$, annual interest rate ($i$) = 0.05, and $n = 3$ years (Table 7.2). The annual repayment to amortize the loan is

$$Z = \frac{1,000(0.05)}{H} = \$367$$

where $H = 1 - (1 + 0.05)^{-3}$.

The following steps are used to construct the table:

- If $n = 3$, construct 4 rows so that for Year 4, the principal at the start of the period should be 0 to amortize the loan.

Table 7.2 Amortizing loan repayment table.

Year	Principal at start of period	Annual repayment	Payment of Interest	Principal
1	1,000	367	50	317
2	683	367	34	333
3	350	367	17	350
4	0			

- Fill in the principal amount ($1,000) and annual repayment ($367) for Year 1.
- For Year 1, Interest payment = Principal at start of period ($1,000) × Interest rate (0.05) = $50. Payment of principal = 367 − 50 = $317.
- For Year 2, Principal at start of period = 1,000 − 317 = $683. Interest payment = 0.05(683) = $34. Payment of principal = 367 − 34 = $333.
- For Year 3, Principal at start of period is 683 − 333 = $350. Interest payment = 0.05(350) = $17. Payment of principal = 367 − 17 = $350.
- For Year 4, Principal at start of period = 350 − 350 = $0.

Balloon loan repayment

Not all loans are amortizing, such as the following repayment schemes for a $70 loan in a week:

1	2	3	4	5	6	7
$10	$10	$10	$10	$10	$10	$10
$1	$1	$1	$1	$1	$1	$64

Assume that there is no interest on the loan to simplify the discussion. In the first scheme, equal repayments of $10 each day reduces the debt to zero by the end of the week. In the balloon scheme, equal repayments of $1 each day reduces the debt to $64 at the end of the week. A balloon loan is useful if the borrower does not have adequate funds to repay $10 a day. For example, the rent may not cover the annual repayment on a 7-year housing loan. The investor or borrower may request for a balloon loan to reduce the annual cash outlay. At the end of the 7th year, he will sell the house and use the proceeds to repay the remaining debt.

The same principle may be applied to an infrastructure project that generates low cash flows during the initial periods. If these cash flows are insufficient to cover periodic debt repayment, a balloon loan may be suitable.

Consider a $1,000 balloon loan ($L$) with an annual interest rate (i) of 5%, a term (T) of 3 years, and an amortization period (n) of 10 years. The difference between T and n is that the loan must be fully repaid by T, and

n is the number of periods used to compute the periodic repayment. The larger the value of n, the smaller is the periodic repayment. If $n = T$, the debt will be zero at the end of the term, as in the previous amortizing loan example. If $n > T$, it is a balloon loan (Table 7.3).

Table 7.3   Loan repayment table for a balloon loan.

Year	Principal at start of period	Annual repayment	Payment of Interest	Payment of Principal
1	1,000	130	50	80
2	920	130	46	84
3	836	130	42	88
4	748			

The annual repayment is

$$Z = \frac{1,000(0.05)}{H} = \$130$$

where $H = 1 - (1 + 0.05)^{-10} = 0.3861$. At the end of the 3-year term, the borrower must pay the remaining $748.

If n is stretched to 20 years, then $H = 1 - (1 + 0.05)^{-20} = 0.6231$. Hence

$$Z = \frac{1,000(0.05)}{0.6231} = \$80.$$

The annual repayment is lower.

Currency risks

Currency risks include project sales, input purchases, and loan repayments in different currencies for different types of loans as well as restrictions on foreign currency transactions. If the project is in a developing country, sponsors may not be able to secure sufficient local currency loans to match its revenues because the financial market is not well developed. This financing gap exposes the SPV (borrower) to currency fluctuations that are difficult to predict.

Sponsors may mitigate the impact by indexing the output price (i.e. project revenues) to the exchange rate. Clearly, this requires the agreement of the purchaser in the off-take contract.

To overcome restrictions on foreign currency transactions, sponsors may set up an offshore escrow account. It is a bank account with a financial institution outside the home country. An escrow is a financial arrangement where the third party (e.g. a bank) holds and regulates the payment of funds between two parties (e.g. SPV and the off-taker). The purchaser or off-taker will pay in hard currency to this account.

Alternatively, the SPV may try to obtain a government guarantee on foreign currency availability.

Currency swap

An alternative strategy is a currency swap. Suppose a project borrows at 6% in a foreign currency (FC), such as the US dollar, but revenues are in the local currency (LC). It can arrange with a swap bank to receive 6% in FC and pay 8% in LC. Its net position is as follows:

Pay lender	6% in FC
Receive from swap bank	6% in FC
Pay swap bank	8% in LC
Net position	8% in LC

Effectively, the SPV is paying 8% in LC, which matches its revenues. It receives 6% in FC from the swap bank to pay its lender.

Currency fluctuations

Predicting currency movements is not easy. The textbook theory is that a country with continuing large *trade* surplus will see an appreciation of the currency because of greater demand for the currency. In turn, the appreciation will cheapen imports and make exports more expensive, which will reduce the trade surplus. This market adjustment is automatic.

More accurately, currency also flows through *services* and inter-government transfers (e.g. foreign aid). The net flow is the current account balance on trade and services. It will be a deficit or surplus.

Currency also flows through *asset* sales and purchases, and the net flow is the capital account balance. In theory, the balance on current and capital accounts offset each other. In practice, there will be some statistical discrepancy trying to estimate a country's sales and purchases with the rest of the world. Hence, if the US is running a current account deficit, more US dollars are leaving the country than coming in. Foreigners will use the excess dollars to buy US assets, such as Chinese purchases of US bonds and gold.

Apart from market adjustment, the government may influence its exchange rate through interest rate adjustment, direct intervention, currency peg, and bond issues. The value of a currency tends to rise or fall with interest rates. For example, if US interest rates rise, holders of foreign currencies may convert their holdings to dollars and deposit them with American banks. Hence, it is easy for a central bank, such as the Central Bank of Malaysia, to change the interest rate to defend the currency. However, interest rate adjustments affect other sectors of the economy.

Direct intervention refers to the buying and selling of a foreign currency in the forex market. For example, Singapore uses its foreign reserves to buy and sell S$ so that it fluctuates within a band. The policy is to have a strong S$ against a basket of currencies of its major trading partners to fight inflation. The downside is that it may incur large forex losses (Subhani, 2023).

Some countries, such as Hong Kong SAR, peg their currencies to the US dollar. If US interest rates rise, the value of the US dollar will rise, and the HK$ will also appreciate. The Hong Kong Monetary Authority (HKMA) will have to sell US dollars to prevent the HK$ from appreciating too much. If US interest rates continue to rise, such as over the 2021–23 period, HKMA will deplete its US dollar reserves (Yiu, 2023).

A Chinese exporter paid in US dollars will need to exchange it for RMB with a Chinese bank to pay his workers, rent, utilities, taxes, and inputs. The Chinese bank then exchanges its dollar holdings for RMB with the Chinese Central Bank. The Central Bank issues bonds to soak up the excess amounts of RMB swirling in the economy, which explains the relatively low inflation despite the huge trade surplus. It then uses part of its dollar holdings to purchase US assets.

Finally, speculation also affects the exchange rate. If speculators and investors sense that a country may devalue its currency because of balance of payments difficulties, they will sell financial and real assets in the local currency and convert the funds into hard currencies. Or they may borrow in the local currency and convert them into hard currencies. Just before the 1997–98 Asian financial crisis, speculators borrowed in Thai baht and converted them to US$. All they have to do next is to wait for the baht to crash to profit from the devaluation.

In summary, the value of a foreign currency fluctuates because of the balance on current account, which tracks the flows of a currency in and out of a country through trade in goods and services, the buying and selling of assets, receipts and payments on various sources of incomes, and inter-government transfers. In addition, governments use various mechanisms to influence the exchange rate to keep their exports competitive, cheapen imports, or control inflation. Finally, speculators can exacerbate a financial crisis.

Inflation risk

Inflation is the decline in the value of a currency in terms of the goods and services one can purchase. The causes of inflation are the demand and supply of goods and services, which makes it hard to predict. Headline inflation is measured monthly using a pre-determined basket of goods and services containing thousands of items. Core inflation removes the more volatile items from the basket, such as energy and food prices. For example, the prices of some food items are volatile because of changing weather patterns. Similarly, the price of oil is susceptible to geopolitical tensions. Presumably, these fluctuations are transitory and, hence, not part of the core.

On the demand side, excessive purchasing power, such as by increasing money supply, will result in a general rise in prices. On the supply or cost side, prices may increase because of rising wages and supply restrictions or disruptions. These disruptions may be caused by labor strikes, natural disasters, war, and so on.

A price rise in one sector may not lead to a general rise in prices because buyers will seek cheaper substitutes. However, some commodities, such as oil, are used in many sectors and have few substitutes. Hence, there is a tendency to associate rising oil prices with inflation.

Inflation affects firms, households, and government. Consumption demand tends to fall with higher prices, sending the economy into a recession. Consumers with low or fixed incomes tend to suffer more from the effects of inflation. Savers will end up with fewer real dollars, but borrowers with fixed repayments, such as a high-indebted government, will repay with cheaper dollars. Workers will often find that their real wages have fallen, which leads to successive rounds of wage bargaining. It can lead to a wage-price spiral.

To mitigate the inflation risk in a project, it is common to index prices to inflation. In energy-intensive projects, there is often a pass-through arrangement linking output prices to fluctuations in input energy prices.

Interest rate risk

Like all commodities, the level of interest rates depends on the demand and supply for loanable funds. The demand for funds depends on expected profitability and the cost of funds. The supply of funds depends on money supply, interest rates, and how bankers feel about business conditions.

Globally, real interest rates tend to follow the US rate because mobile speculative funds or "hot money" flows to financial assets that earn higher risk-adjusted interests. These flows will affect the exchange rate. For instance, when the Bank of Japan kept interest rates low in the first half of 2023 despite rising US interest rates, investors sold the Japanese Yen and converted them to dollars, and the Yen tumbled.

One way to mitigate interest rate risk is to negotiate for a fixed rate loan that will be higher than a floating rate loan to compensate the lender for the risk. In a floating rate loan, the interest rate is adjusted periodically, such as once every three months. The adjustment is based on a bank's internal rate or the industry benchmark rate. The fixed deposit rate is an example of an internal rate. The inter-bank rate is an example of an industry standard. For instance, the bank may quote a fixed rate of 5% and a floating inter-bank rate + 2%. If the inter-bank rate is currently 2%, then a borrower is paying 2% + 2% = 4%. Apart from the higher cost of funds, a fixed rate loan will be more expensive if interest rates fall. If the inter-bank rate falls to 1%, the borrowing cost is only 3%. Hence, a fixed rate loan is preferable if the borrower expects the interest rate to rise in future.

However, the lender may have the same expectation. Hence, the lender will either not provide fixed rate loans or raise the chargeable interest rate sufficiently high to compensate for the risk. By offering floating rate loans, lenders shift the interest rate risk to borrowers. It may be possible for the borrower to negotiate for the flexibility to convert from floating to fixed rates and vice versa.

In collared financing, the interest rates fall within a range, say 5–10%. The borrower pays no more than the upper limit. However, if interest rates fall below the lower limit, the lender will charge the lower limit.

Interest rate swap

Sponsors can also arrange for an interest rate swap with a swap bank or party to mitigate interest rate risks. Suppose the SPV borrows at S + 2%, where S is the Singapore inter-bank rate. It may arrange to swap it with another bank by paying a 3% fixed interest rate and receiving S%. The net position of the SPV is as follows:

Pay to lender	S + 2%
Pay to swap bank	3% fixed
Receive from swap bank	S%
Total payment	S + 5%
Total receipt	S%
Net position	5% fixed

The fixed rate is also called the swap rate. Most banks offer interest rate swaps, swapping floating rates for fixed rates and vice versa.

Financial feasibility

There are three main financial tables for a firm, namely,

- Income Statement;
- cash flow table; and
- Balance Sheet.

Financial ratio analyses based on the above tables are used to check the SPV's liquidity, efficiency, profitability, and solvency.

Income Statement

The income or profit and loss statement shows the profit or loss for a particular year or over several years (Table 7.4). As the PPP contract has not yet been awarded, the figures in the sponsor's Income Statement are projected values. The figures in the table are for illustration.

The cost of goods sold are factory or site costs. They include production cost of materials, labor, factory overhead, and transport. If desired, these costs can be reflected as line items.

The selling, general, and administrative expenses (SGA) are head office costs. They include marketing, sales, management, insurance, legal expenses, secretariat, office supplies, utilities, and office equipment. Research and development is booked as a separate line item.

EBITDA stands for earnings before interest, tax, depreciation, and amortization. Investment is booked as an annual depreciation rather than as an initial lump sum to better reflect profit from operation. The allowable

Table 7.4   Income Statement ($m).

	Year 1	Year 2, etc.
Revenue	30	
Less: Cost of goods sold	15	
Gross profit	**15**	
Less: Selling, general and administrative expenses	5	
Research and development	0	
Operating profit (EBITDA)	**10**	
Less: Depreciation and amortization	5	
Operating income (EBIT)	**5**	
Less: Interest expense	1	
Corporate tax	1	
Net profit (Net income)	**3**	

depreciation depends on tax laws. If straight-line depreciation is used, the annual depreciation of an asset is

$$A = \frac{I - S}{T}.$$

Here, I is the initial investment, S is the salvage (scrap) value, and T is the allowable depreciation period. For example, if an asset costs $10 m with a salvage value of $2 m and the allowable period is 8 years, then

$$A = \frac{10 - 2}{8} = \$1 \text{ m.}$$

The SPV will have different assets, and needs to work out the depreciation allowances separately for each asset. Land cannot be depreciated because it is assumed to be indestructible. The term "amortization" in EBITDA refers to the depreciation of intangible assets, such as trademarks, patents, copyright, and goodwill. It should not be confused with the amortization of a loan in Table 7.2.

The net profit is obtained after deducting depreciation, interest expense, and corporate tax. Interest is an expense if the company is paying off a loan.

There are some items that are not reflected in Table 7.3, and they should be reported separately. For example, if the company puts its cash as deposits, it earns interest income. Thus, interest is not necessarily an expense. Further, there may be extraordinary charges or credits such as losses due to disasters, gain or loss from a law suit, or a windfall gain by winning a lottery. If these gains or losses are large, they may distort the income statement. Hence, these items should be reported separately.

Cash flow statement

The cash flow statement shows cash inflows and outflows for the SPV or a company from Year 0 to n (Table 7.5).

The cash flow from operations are taken from the previous Income Statement. Depreciation and amortization are then added because they are non-cash items. This means that the SPV does not actually pay for the depreciation; it is just an accounting rule for tax purposes. It is assumed

Table 7.5   Cash flow statement ($m).

	0	1	...	n
Cash flow from operations				
Net profit	3			
Depreciation and amortization	5			
Change in working capital	0			
Cash flow from investing activities				
Sale/purchase of property, plant, and equipment	−10			
Sale/purchase of long-term financial assets	0			
Sale/purchase of patents, trademarks, and licenses	0			
Cash flow from financing activities				
Proceeds from bank loans	10			
Repayment of funds borrowed	0			
Sale/purchase of marketable securities	0			
Equity issued	2			
Dividends paid out	−1			
Net cash flow	**9**	**10**		
Cash balance at the beginning of the period	0	9		
Cash balance at the end of the period	9	19		

that when a company buys a machine, it pays for it upfront as a book value and then its value depreciates each year. The change in working capital (or net working capital) has been discussed in Chapter 3.

For investing activities, every sale of assets produces cash, and every purchase reduces cash. In this example, the firm spent $10 m to purchase property, plant, or equipment.

For financing activities, the firm received $10 m as a loan, issued $2 m of equity, and paid out $1 m in dividends.

The net cash flow is the sum of cash flows from operations, investing activities, and financing activities. Finally, we may compute the cumulative cash balance for each year.

Analysts use the term *free cash flow* (FCF) to determine if a firm has freed-up cash; that is,

FCF = EBIT – Tax + (Depreciation and amortization) – Capex – ΔNWC.

The first three terms are taken from the Income Statement. Capex refers to planned capital expenditure, and ΔNWC is the change in net working capital.

In summary, there are three ways of looking at a firm's cash flows. EBITDA does not include capex, FCF does not consider financing activities, and the cash flow statement in Table 7.5 is the most complete but requires more effort in gathering the data.

Equity internal rate of return

Table 7.5 is the cash flow statement of the *firm*. It has to be modified for *project* analysis. Sponsors are interested in the return on their investment in a project, which is equity. Hence, they are interested in the project cash flows.

Suppose a project incurs an initial equity of $30 m and has positive cash flows thereafter:

Year 0	Year 1	Year 2	Year 3	Year 4
–$30 m	$10 m	$10 m	$10 m	$10 m

The equity internal rate of return (IRR or EIRR) (q) for the project is found by solving

$$0 = -30 + \frac{10}{1+q} + \cdots + \frac{10}{(1+q)^4}.$$

Using trial and error, $q = 12.6\%$. It is worth repeating that the figures in the table are cash flows and not project benefits and costs. Unlike the public agency, the private sector (SPV) is interested in profitability, not economic benefits and costs.

Balance Sheet

The Balance Sheet shows a firm's financial health at a particular date, such as 31 December 20xx (Table 7.6). It uses accrual accounting; that is,

Table 7.6   Balance Sheet.

Assets ($m)		Liabilities ($m)	
Current assets		**Current liabilities**	
Cash and marketable securities	20	Accounts payable	14
Accounts receivable	50	Dividend and taxes payable	6
Inventories	40	Short-term loans	10
Total current assets	110	Total current liabilities	30
Long-term assets		**Long-term liabilities**	
Land and buildings	50	Bank loans	20
Machinery and equipment	30	Bonds	30
Less accumulated depreciation	(10)		
Intangible assets	10		
Less accumulated amortization	(5)		
Total assets	185	**Total liabilities**	80
		Net worth (shareholders' equity)	105

- revenue is recognized in the period in which the good is sold or a service is performed irrespective of whether the firm is paid; and
- the accrual principle also applies to the expense side to match expenses to revenues.

Current assets and liabilities refer to short-term items of less than a year, such as cash and marketable securities, accounts receivable, and inventories of goods and materials. Long-term assets include buildings, machinery, and equipment. These assets need to be depreciated except for land. Depreciation for intangible assets, such as patents and trademarks, is called amortization. Hence, depreciation and amortization are often abbreviated as D & A.

On the liability side, the firm has short-term liabilities that it needs to pay out within a year, such as accounts payable, dividends and taxes payable, and short-term loans. It also has long-term liabilities, such as bank loans and bonds. The difference between assets and liabilities is the firm's net worth or shareholders' equity. It comprises common stock paid in by shareholders and retained profits. The Balance Sheet must balance; that is,

$$\text{Total assets} = \text{Total liabilities} + \text{Net worth.}$$

Ratio analysis

Lenders and investors routinely use ratio analysis to assess the performance of a firm. The key ratios are

- liquidity ratios;
- efficiency ratios;
- profitability ratios; and
- solvency ratios.

Most of the figures for the ratios are taken from the Income Statement or Balance Sheet. In Table 7.7, variables with an asterisk are taken from the Income Statement; the rest are from the Balance Sheet. Within each category, there are many ratios, and some examples are given in the table below.

Table 7.7   Examples of financial ratios.

Ratio	Examples
Liquidity	Current ratio = Current assets/Current liabilities Cash ratio = (Cash + marketable securities)/Current liabilities
Efficiency	Inventory turnover = Cost of goods sold*/Average inventory Asset turnover ratio = Revenue*/Total assets
Profitability	Gross profit margin = Gross profit*/Revenue* Return on assets = Net profit*/Total assets
Solvency	Debt to equity ratio = Total liability/Total debt Interest coverage ratio = Operating profit*/Interest expenses*

These ratios are compared with those from similar firms in the industry or with rules of thumb. Analysts also track changes in these ratios over time. Any substantial deviation warrants some explanation; for example, a firm may have a much higher debt to equity ratio than the industry average, signaling excessive borrowing.

Example: China's three red lines

In August 2020, China introduced the three red lines to strengthen the financial position of property developers amid rampant property speculation and the housing bubble. The red lines are:

- liabilities should not exceed 70% of assets;
- net debt should not exceed 100% of equity; and
- money reserves must be at least 100% of the short-term debt.

One such developer that found itself on the wrong side of the red lines is Evergrande Group, one of the largest developers in China. Initially, it borrowed from the domestic market but soon went to issue offshore bonds. In August 2023, it filed for US bankruptcy protection.

Project insurances

As part of project preparations, sponsors have to consider insurance. The basis of insurance is often claims-made; that is, claims must be made before the expiry of the policy. There is also excess, an amount below which the policy holder has to foot the bill. For example, if there is an excess of $1,000 on a car insurance, the insurer will not pay for any damage below this amount. This is to encourage insured drivers to be careful and avoid frivolous claims. There are also monetary cover limits beyond which a policy holder cannot claim.

Political risk insurance (PRI) provides cover against political events that lead to loss, such as terrorism, expropriation, civil unrest, and so on (see Table 3.3). Such insurances may be purchased from multilateral development banks or private insurers. For example, the World Bank's Multilateral Investment Guarantee Agency (MIGA) provides PRI against foreign exchange restrictions, expropriation, war and civil disturbance, and breach of contract by the government. Premiums for PRI vary depending on the insurer, insured sum, and country risk. Obviously, the latter tends to be higher for developing countries.

Work injury compensation insurance is usually a statutory requirement for all employees or those earning below a certain salary. It covers injury or death on a no-blame basis; that is, an injured worker of a contractor will be compensated even if he is at fault. However, if a worker feels that the payout is not sufficient, he may forgo it and sue the employer instead. Employers may want to insure against this possibility by purchasing *employer's liability insurance*. In addition, the contractor needs to insure his property, vehicles, machinery, and other assets.

It is common for the contractor to purchase *builder's risk insurance* and include the premium as part of the bid. In some projects, the sponsor (owner) purchases the insurance, but he or she may not be familiar with the policy or claims. The sum insured is usually the estimated project value and covers project works, temporary structures, and materials (on-site, in transit or stored off-site) against named perils such as water and fire damage, theft, and mistakes. Normally, it excludes major catastrophes, damage to project documents, reworks, and faulty design. In some cases, it may be possible for the local insurer to re-insure catastrophic risks with an international insurer who can diversify the risks across many cities on a global scale. Sponsors have to conduct their due diligence to ensure that the contractor obtains adequate insurance without too many qualifications.

A *general liability insurance* (GL insurance) protects the insured against third-party claims for property damage or injury. For example, a crane may damage an adjoining property or a pedestrian is hit by falling debris. The contractor either purchases GL insurance separately or includes it as part of the builder's risk insurance, which then becomes a *builder's all-risk insurance*. The term is misleading because there are exclusions, as discussed above. Normally, the sponsor is included as an *additional insured* party because the neighbor whose property is damaged may sue the sponsor and contractor. The sponsor, as an additional insured party, may file a claim to cover the loss.

It is possible to include *environmental liability insurance* to cover clean-up cost under the builder's risk insurance. If it is not possible, the contractor has to purchase separate insurance under similar arrangements.

Sponsors need to ensure that consultants have professional indemnity insurance, also called *professional liability insurance*. It covers consultants' inadequate advice, services, or designs due to errors and omissions (mistakes) or negligence (carelessness) that cause the SPV to lose money. If the SPV sues the consultant, the insurance should cover the legal costs of the consultant and any award to the sponsor.

If the project is delayed, the SPV may suffer from a *consequential loss* of revenue. It is possible to purchase insurance for start-up delay or rely on liquidated damages from the contractor to cover the loss.

Finally, sponsors need to purchase *operation and maintenance insurances*. These insurances include

- operator's all risk insurance to cover property damage, injury, and general liability to third parties;
- machinery breakdown; and
- business disruption.

Since the SPV has a project loan, the lender will require that the benefits of insurance be assigned through *lender's clauses* in the policy. If the SPV defaults on the loan, the lender has the right to the benefits.

Over the years, insurers have developed new instruments to insure against different types of project risks. An example is a catastrophe bond, as explained in the following example.

Example: Catastrophe bond

If a project such as a hydropower plant is to be located in an earthquake zone, it may not be able to find a local insurer with the capacity to insure against an earthquake. Sometimes, the local insurer may take the risk and reinsures with an international insurer. The latter is better able to spread the risks globally; after all, earthquakes do not happen all at once in different places around the world.

Another possible solution is for the insurer to issue catastrophe bonds, or cat bonds, to investors. The insurer sets up a special purpose entity (not to be confused with the project SPV) to

- receive premiums from policy holders (e.g. the infrastructure project company);
- issue cat bonds to investors; and
- invest the premiums and investors' funds for stable returns.

The cat bond has a short duration, such as three years. If the earthquake does not happen, investors get back their principal in addition to periodic coupons. Cat bond coupons pay higher interest to compensate investors for the higher risk.

If an earthquake occurs as defined by pre-determined triggers (e.g. above a certain value on the Richter scale), investors will get back part of the principal depending on the extent of damage. The insurer will recall the invested funds and use them to pay policy holders for the earthquake damage. The remaining unused funds will be returned to cat bond holders. To incentivize investors, the insurer is likely to limit the insurance payouts and protect part of the principal.

The World Bank (WB) also issues cat bonds for various purposes, such as for the protection of the Government of Jamaica (GoJ) against tropical cyclones. The structure (Fig. 7.1) is as follows:

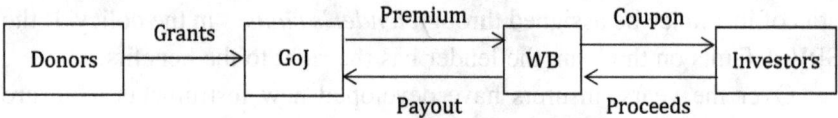

Fig. 7.1.   Example of a cat bond issue.

The coverage is US$185 m over 2.4 years (July 2021 to December 2023), with a coupon of 4.4%. GoJ uses the grants from international donors to pay the premium, and will receive a payout from the World Bank if a cyclone above an agreed level of severity occurs.

Draft shareholders' agreements

At the end of the bid preparation stage, sponsors take into consideration all the issues and draft the shareholders' arrangements and project agreements (Table 7.8). Subsequently, lenders will draft the loan and security documents, Shareholders' Support Agreement, and Direct Agreement (see Chapter 8). They will also provide inputs or require changes to other agreements to protect the lenders' interest. The grantor will draft the PPP contract or the off-take contract (see Chapter 5). If the buyer of the project output is a private entity, the SPV will draft the off-take contract.

We have dealt with the *Development Agreement* at the start of this chapter. The *Shareholders' Agreement* covers

- the establishment of the SPV;
- shareholding, funding, and management;
- additional equity for cost overrun;
- bank guarantee on obligations;
- dividend policy, in-kind contribution, and disposal of shares;

Table 7.8   Project contracts and agreements.

Shareholders' agreements
Development Agreement
Shareholders' Agreement
Shareholders' Support Agreement
Loan and security documents
Loan Agreement
Security Agreement
Equity Support Agreement
Common Terms Agreement
Accounts Agreement
Inter-creditor Agreement
Project agreements
PPP Contract
Direct Agreement
Construction Contract
O & M Contract
Off-take Agreement
Fuel Supply Agreement
Land Lease Agreement
Equipment Contract

- conflict of interest, such as if the contractor or operator is also a shareholder;
- undertakings not to engage in activities in competition with the SPV; and
- dispute resolution.

Lenders may require a *Shareholders' Support Agreement* that covers

- technical support;
- further equity contributions;
- restrictions on disposal of shares; and
- completion, cost overrun, and liquidated damages guarantees.

Draft project agreements

The draft project agreements are briefly outlined below.

Direct Agreement

The grantor, lender, and SPV sign a Direct Agreement to allow the lender to *step in* if the SPV defaults on its loan. This tri-partite arrangement will require negotiations among the parties. The lender will try to "cure" the project if the default is temporary. The cure period may last up to three years. If the project is in serious trouble, lenders may replace the management team and novate the project agreements. The lender will also step in to secure project assets and agreements over third-party claims, such as that of the contractor, subcontractors, and suppliers.

The grantor may step in if the project suffers from serious performance lapses, threatens national security, or raises serious environmental concerns or public health. For example, it may step in in the public interest if a water desalination project does not deliver sufficient quantities of water. The grantor may compensate the SPV for the present value of the asset and take over its operations.

Some countries do not have step-in rights for lenders or grantors. In countries with step-in rights, lenders have reservations on the ability of the grantor to step in because of a possible loss on the loan. There is also the question of who steps in first if the SPV defaults on the loan as well as on its obligations to the grantor.

Construction Contract

The clauses in the Construction Contract will depend on the procurement strategy. Lenders prefer traditional fixed price contracts for building projects to mitigate risks of price escalation. For process projects, the EPC contract is preferred to provide a single point of responsibility. The standard conditions of contract for building works are given below as an example:

- Definitions, interpretations, and applicable laws;
- Consultants' representatives;

- Contract documents;
- General obligations of the contractor;
- Sub-surface and ground conditions;
- Contractor's design responsibility;
- Acceleration of works;
- Liquidated damages;
- Substantial completion;
- Quality standards and defects;
- Variation of works;
- Notices and fees;
- Site surveys;
- Project program;
- Extension of time;
- Site possession and commencement of works;
- Suspension of works;
- Measurement of works;
- Claims;
- Construction plant and equipment;
- Temporary works, materials, and goods;
- Indemnity provisions;
- Insurances;
- Subcontractors;
- Termination;
- Progress payment;
- Final account;
- Fluctuations;
- Final completion; and
- Dispute resolution.

In some projects, there may be special conditions such as incentive payments for early completion, advance payment for materials or mobilization, assignment of rights to lenders, early occupation of the site, and so on. These conditions are consolidated under a separate document called "Special conditions of contract."

There are many standard forms of contracts drafted by civil engineers, government agencies, developers, architects, and international bodies (Haswell and de Silva, 1989; Ramus *et al.*, 2006; Lim, 2020). These forms may represent particular interests and need to be used with care.

Operation and Maintenance Contract

The Operation and Maintenance (O & M) Contract may contain the following provisions:

- contract period;
- scope of services comprising
  - operations, including training of operatives;
  - maintenance, that is, routine maintenance, corrective maintenance (repairs), preventive maintenance (regular inspection and servicing), and cyclical works; and
  - other services such as participation in commissioning tests, security, emergency, waste management, energy conservation, renovations, and leasing.
- performance-based fees;
- performance standards;
- limits on the operator's authority regarding disposal of assets, contracting, and reimbursable expenses not in conformity with the budget;
- operator's assistance in obtaining statutory permits and approvals;
- guarantee of compliance with the law;
- operator's insurances;
- indemnity;
- monthly and annual progress reports;
- SPV's right to inspect facilities and operations;
- SPV's right to inspect and audit records;
- early termination; and
- dispute resolution.

The O & M Contract will also normally stipulate the owner's responsibility to

- provide manuals, list of spares, and drawings;
- provide feedstock, site access, and utilities;

- approve budget and operating plans;
- pay reimbursable costs;
- engage external parties for major works beyond the scope of services provided by the operator; and
- develop a standard operating procedures (SOP) manual on reporting, accounting, contracting, and record-keeping.

Off-take Agreement

The Off-take Agreement covers

- type of purchase contract, preferably "take or pay" where the buyer has to pay for a minimum amount of the output even if he does not take delivery;
- payment mechanism;
- adjustment for inflation and currency movements, if any;
- pass-through provisions, if any;
- penalties for poor performance;
- inter-connection facilities;
- transport and storage requirements;
- termination;
- *force majeure* provisions; and
- dispute resolution.

Supply Agreement

There are many types of supply and purchase agreements depending on the role of the SPV in the supply chain. For example, if the SPV owns the pipeline that transports natural gas, it signs a *Through-put Agreement* to transport the gas at a specific price on a monthly basis. In contrast, if the SPV processes the raw materials, it signs a *Tolling Agreement* to charge the counter-party for the processing.

A *Fuel Supply Agreement* is often a "supply or pay" contract with a credit worthy and reliable supplier. The credit worthiness is required in case the supplier needs to pay liquidated damages. The agreement will stipulate the supply price, range of quantities, and interconnection facilities provided by the supplier. Both sides will need to back their commitments, such as with Letters of Credit.

Other Agreements

A *Land Lease Agreement* stipulates the appropriate land use, term of lease, rent, payable tax, renewal options, responsibilities for repairs and maintenance, and condition of the land at the end of the lease. The owner, usually the government in PPP projects, will want to retain the right of access.

An *Equipment Contract* for major equipment often contain the following provisions:

- specifications;
- price, taxes, and import duties;
- milestone dates;
- delivery arrangements, including insurance, transport, and handling charges;
- installation, inspection, commissioning tests, certification, completion, and taking over of ownership;
- possible defects, acceptance, payment, retention, and liquidated damages;
- warranty;
- maintenance and repair contract (if any), including schedule of rates;
- training of operatives;
- manuals and list of spares; and
- dispute resolution.

Business case

The last step in the bid preparation is to write the business case to seek parent company approval and secure project financing. The business will cover most of the items discussed in this chapter, such as

- project information;
- shareholding structure;

- project governance structure;
- design proposal;
- price proposal;
- political, regulatory, and legal assessments;
- projected revenue;
- cost and schedule estimates;
- quality measures;
- inputs;
- O & M;
- *force majeure* risks;
- hedging instruments;
- project insurances;
- draft shareholders' agreements;
- draft project agreements; and
- financing strategy.

The financing strategy will include proposed equity contributions and the amount of loan requested.

The information is confidential, and sponsors will have to decide what to include and exclude when presenting the business case to lenders.

References

Berle, A. and Means, G. (1932) *The modern corporation and private property.* Livingston: Transaction Publishers.

Durand, D. (1952) *Cost of debt and equity funds for business.* Cambridge: NBER.

Fair, R. (1972) Disequilibrium in housing models. *Journal of Finance,* **27**(2), 207–21.

Haswell, C. and de Silva, D. (1989) *Civil engineering contracts.* London: Butterworth.

Joshi, M. (2010) *The concepts and practice of mathematical finance.* London: Cambridge University Press.

KPMG (2010) *PPP procurement: Review of barriers to competition and efficiency in the procurement of PPP projects.* Sydney: KPMG.

Lim, P. (2020) *Contract administration and procurement in the Singapore construction industry.* Singapore: World Scientific Publishing.

McDonald, R. (2014) *Derivatives markets.* Harlow: Pearson.

Ramus, J., Birchall, S., and Griffiths, P. (2006) *Contract practice for surveyors.* London: Butterworth-Heinemann.

Subhani, O. (2023) MAS books largest loss of $30.8b in effort to fight inflation. *The Straits Times*, 5 July.

Tan, W. (2022) *Research methods: A practical guide for students and researchers.* Singapore: World Scientific Publishing.

Yiu, E. (2023) Hong Kong steps into market for the first time this year, selling US$538 million to prop up local dollar as funds flee to higher interest rates. *South China Morning Post*, 14 February.

CHAPTER 8

Bid evaluation and contract award

Price bids

As discussed in Chapter 5, the grantor evaluates bids for infrastructure projects on design, price, and other criteria on a weighted basis. The weights, which tend to be ad hoc, depend on the grantor's priority.

To begin, we consider only price bids, tenders, or auctions using the highest or lowest price. In this chapter, the terms bid, tenders, and auctions will be used interchangeably. If the grantor is the buyer (off-taker) of the project output, such as in desalination projects, the bid is based on the lowest price. If the grantor is the seller, such as in mobile spectrum auctions, the award is based on the highest price. To simplify the discussion, we will consider bidding based on the highest price, as the principles are easily extended to bids based on the lowest price.

Let H be the highest bidder at \$100 and S the second highest bidder at \$90. These empirical values are for illustration, and are known to only H and S, respectively. Bids are progressively raised at a minimum of \$1 increment.

There are many types of auctions (Klemperer, 2014; Krishna, 2009), and the common ones are discussed below.

English auction

The English auction uses open bidding; that is, every bidder has the chance to see what others are bidding. It starts with a floor or reserved price, and bids are progressively raised until there is only one bid left, the highest bid. *H* pays slightly above the second highest valuation because the bidding stops at $91.

Japanese auction

The Japanese auction is similar to the English auction. However, the price increase is not based on the next higher bid. It is automatic, such as $1 per second, until only a single buyer remains in the auction room.

Similar to the English auction, *S* will exit the room when the bid exceeds $90 and the winning bid is also $91.

Dutch auction

In a Dutch auction, open bidding starts with a high price set by the seller, and this is progressively lowered until someone raises her hand. This bidding process is often faster than the English auction because it takes only one bid to decide the outcome.

Again, the winning bid is $91. *H* will not raise her hand at $100 because it yields no profit. If she expects a 10% profit, then bidding should stop around $90.

Vickrey auction

The Vickrey (1961) auction uses sealed bids. Bidders do not know what the others are bidding until the envelops are opened. The twist is that the winning bidder pays not what she bids but the second highest bid.

The optimal strategy for *H* is to bid at valuation (i.e. $100) because she will pay the second highest bid, at $90.

Sealed bid auction

Finally, consider a sealed bid auction where a bidder pays what she bids. Again, *H* will not bid $100 because it yields no profit. Hence, she will bid around $91, which is slightly higher than what she thinks is the second highest bid.

Revenue equivalence

In summary, the English, Japanese, Dutch, Vickrey, and sealed bid auctions give the same result; that is, the winner pays the second highest valuation. This is called the revenue equivalence theorem; that is, if bidders are risk-neutral, the different types of auctions give the same outcome and revenue (Myerson, 1981).

But is it worth it? We have to consider the winner's curse.

Winner's curse

The winner's curse is the difference between a winning bid and the true worth of an item. The difference between the winning and second bid is not a winner's curse but probably a winner's regret.

Over-bidding can arise if a bidder is emotional or desperate or if the bidding environment is intense and competitive. It can also occur for strategic reasons, such as to gain experience, enter a market, or deter competitors from entering the market. A winner's curse need not always exist. A bidder may have private information about the value of the item that is not available to other bidders as distinct from common value auctions where bidders do not have such private information. This brings us to how imperfect and asymmetric information may lead to over-bidding.

Over-bidding

For infrastructure projects, a major cause of over-bidding is information imperfection or asymmetry. Information is imperfect if it is not fully known. The revenues, costs, and risks are only estimates and not fully known. Hence, there is the possibility of forecast errors when submitting a bid (Standard and Poor's, 2002).

Information is asymmetric if one party has better information than the other. In this case, the grantor has better information about the profitability of the project than the SPV (i.e. bidder). For example, the grantor may paint a rosy picture about the profitability of the project to attract bidders.

Sponsors may mitigate these risks by doing their due diligence (Ferris and Petitt, 2002) or bid more conservatively. The winner's curse may also be smaller if there is an opportunity to renegotiate the PPP contract early in the concession period (Guasch, 2004).

Design and price bids

For price and design bids, the grantor awards the contract based on design (technical) and price. As discussed in Chapter 5, the grantor will reject bids with deficient designs; that is, those with critical deviations from specifications, as well as bids below a certain design score. A variation of this process is to allow bidders to make minor adjustments to the design (Asian Development Bank, 2018).

In the second stage, the grantor may use a points system to award the contract based on design and price.

Best-value bids

Best-value bids are evaluated on design, price, and other criteria such as financial strength, innovation, track record, environmental sustainability, operation and maintenance regime, and schedule. Usually, a simple points system is used.

Issues in bid evaluation

There are a number of issues in evaluating bids, such as

- having one bidder only;
- no conforming bids;
- abnormally low bids

The grantor is not obliged to award the contract if the responses are not favorable. If there is only one bidder, it is possible to award the contract after conducting internal due diligence or to repackage it and re-tender. An abnormally low bid may signal misreading or misunderstanding of the project requirements, mismeasurement, desperation for work, or the efficiency of the bidder. It is rare for an efficient bidder to submit very low bids.

A second set of issues concerns

- price adjustments for minor technical deviations;
- admissibility of alternative designs; or
- bids with qualifications that include or exclude certain items.

For bids with price adjustments for minor technical deviations, the practice varies, but it is common to allow for them because the technical deviations are not major.

Bidders may suggest changes to improve the design or to include or exclude certain items. The grantor should welcome such suggestions to improve the design. Normally, the tender conditions stipulate the admissibility of such suggestions, but the grantor will award the contract based on the original design.

Bid opening

For PPP projects, bid opening is public in the interest of transparency. All bidders are then notified of the results.

Award of contract

The grantor will sign a Letter of Intent (LoI) with the winning bidder to enter into a PPP contract. The PPP contract is not signed until the winning bidder has reached financial close with the lender or until the grantor has finalized the site. Depending on the complexity of the project and the financial environment, it may take up to a year to reach financial close.

The contract clock starts when the grantor provides the winning bidder with a Notice to Proceed.

References

Asian Development Bank (2018) *Guide on bid evaluation*. Manila: ADB.

Ferris, K. and Pettit, B. (2002) *Valuation: Avoiding the winner's curse*. New Jersey: Prentice Hall.

Guasch, J. (2004) *Granting and renegotiating infrastructure concessions: Doing it right*. Washington: World Bank Institute.

Klemperer, P. (2014) *Auctions: Theory and practice.* New Jersey: Princeton University Press.

Krishna, V. (2009) *Auction theory.* New York: Academic Press.

Myerson, R. (1981) Optimal auction design. *Mathematics of Operations Research,* **6**(1), 58–73.

Standard and Poor's (2002) *Traffic forecasting risk in start-up toll facilities.* London: Standard and Poor.

Vickrey, W. (1961) Counter-speculation, auctions, and competitive sealed tenders. *Journal of Finance,* **16**(1), 8–37.

CHAPTER 9

Lender's due diligence

In-principle approval

Once the sponsors decide to bid for a project, the financing process gets underway. They will explore the market to determine the best financing package among groups of syndicated lenders. Sponsors may approach certain lenders where they have previous business dealings or issue a Request for Proposal for lenders to bid.

Lenders conduct due diligence before issuing in-principle approval to fund the project if the borrower (sponsor) wins the PPP contract. The due diligence is required because of information imperfection and asymmetry; that is, profitability is based on forecast revenues and costs, and borrowers may hide or underestimate the real project risks to secure funding at a lower cost or with better terms. Raising the interest rate may not deter risky borrowers, and it increases the risk of default. Other than due diligence, lenders reduce their risk exposure by forming loan syndicates to share the risks and by lowering the quantum of loan. Collectively, these tactics are called non-price measures (Harris, 1974).

After obtaining the Letter of Intent from the grantor, sponsors have about three to six months to reach financial close. The duration depends on the size and complexity of the project. In some cases, the period may be extended, such as during a financial crisis when credit is tight.

In an underwritten deal, the lead arranger or lead bank of the loan syndicate may provide firm commitment and underwrite the entire loan. It will absorb the difference if the syndicate is unable to raise the loan amount. It will subsequently try to sell the remaining loan separately, possibly at a discount. Such a deal is possible if the lender is confident about the project.

For riskier projects, the lender may seek to underwrite only part of the loan, called a "best-efforts" deal. The lender will put in the "best efforts" to market the loan to possible lenders.

Finally, in a "club deal," the lead arranger does not underwrite the loan. Instead, a club or group of banks issue the loans together. Each lender will hold on to her portion of the loan without exiting. The borrower may initiate a club loan by being its arranger and then deal with different banks. The loan documentation is identical for all participating banks.

Borrower's background

Lenders begin their due diligence by studying the borrower's background, business model, track record, financials, and resources. The borrower must have a sound business strategy, and the project's goals and objectives are aligned with the strategy.

Project information and structuring

The project information and structuring are obtained from the business case report discussed in Chapter 7.

Expert panel

Lenders may use a panel of independent experts to assist in the due diligence. They include the market analyst, project expert, technical expert, and possibly a common or lead counsel. Each lender will have her own legal counsel. Depending on the type of project, lenders may appoint other experts for opinions on environmental assessment or a local legal counsel familiar with the laws of a developing country.

The market analyst provides a report on the reasonableness of the sponsors' market analysis. For large private projects, lenders require pre-lease commitments as evidence of good demand before approving the loan. For PPP projects, pre-construction off-take commitments serve a similar purpose.

The project expert provides inputs on all aspects of project management, such as the scope, preliminary schedule, cost estimates, design management, expected quality, procurement strategy, and project execution.

The technical expert provides an independent review of the specifications, scale, and technology. During the construction phase, the technical expert provides inputs on technical issues that may crop up.

The common or lead counsel of the syndicate provides inputs on the draft project contracts and agreements by the sponsors (i.e. borrower) and other legal opinions. If the project is in a developing country, there are likely to be legal issues regarding contracts, property rights, and their enforcement. The lead counsel should set forth the basis of her actions with the counsels representing the other lenders through a prior representative letter (Ryan, 2009).

From these inputs, lenders may ask for adjustments to the financial model, such as changes to demand projections, costs, schedule, and even the basic engineering technology. If both parties are satisfied with the proposed changes, the next step is to negotiate the terms of the loan.

Loan negotiation

The following terms are likely to be negotiated:

- quantum of loan;
- equity;
- use of subordinated debt (shareholders' loans) as part of equity;
- loan profile;
- sponsors' contingent equity support for cost overrun;
- restrictions; and
- default triggers.

The maximum amount of loan depends largely on the ability of the project to generate sufficient operating profit to cover debt repayment. This ratio should be greater than 1, and lenders often require at least 1.2. Lenders will roll up interest during construction as part of the loan quantum.

Generally, infrastructure projects are financed with about 10 to 40% equity. The use of third-party equity, such as from pension or infrastructure funds, is not uncommon in project finance. Such funds may be treated as convertible debt. Sponsors usually provide subordinated debt rather than equity to reduce their taxes. These shareholders' loans are often considered as equity. Lenders will want to ensure that interest rates on such loans are reasonable and payable only after lenders give their permission for the payment of dividends to shareholders. Usually, this occurs only after the first repayment of the project loan, and the debt-service cover ratio is healthy. Ultimately, lenders do not want sponsors to extract profits from the project in the form of interest on shareholder loans or dividends too early, before the project stabilizes its revenues and operations.

The loan profile contains the mechanics of the loan provided in the Loan Agreement. It includes the first draw, the periodic draws, grace period, and first repayment. The loan is often available for the first draw after the financial close. The periodic draws may be monthly or quarterly. Monthly draws will coincide with the contractor's request for progress payment, while quarterly disbursements will reduce the paperwork. During the grace period, there is no repayment or just the repayment of interest. Upon the completion of construction, the project takes several months or even up to a year to ramp up to full capacity. During this period, the project is not generating stable revenue, and there is usually a grace period.

Sponsors will be required to provide contingent equity support for cost overrun. Normally, the loan does not cover cost overrun, although this may be negotiated. This is a reason why scope creep, variation orders, and other factors that raise project costs cannot be taken lightly.

There will be restrictions on the borrower, such as profit distribution (i.e. declaration of dividends), further borrowings, capital investment, leasing, and sale of shares. This is to ensure that the borrower has the ability to repay the loan.

The final item to negotiate is default triggers set out in the Common Terms Agreement. As discussed in Chapter 7, one consequence of default is the right of lenders to step in to cure the project. Lenders may replace the management if the project is in serious trouble, which is not in the interest of sponsors.

After negotiation, the agreed terms will be documented in various financial agreements discussed below (Khan and Parra, 2003). Further changes after the signing of these documents will be recorded in a separate Amendment and Waiver Agreement.

Financial agreements

The various financial agreements in a PPP project are discussed below.

Loan Agreement

The Loan Agreement contains the loan schedule and reserved discretions. The loan schedule includes the principal, term of loan, amortization period, rate of interest, fees, disbursement conditions, and interest adjustments.

Reserved discretions are items that require lenders' approval. They include

- termination or transfer of any project agreement;
- decisions to initiate arbitration;
- demands for payment, e.g. payment to the contractor;
- assignment of rights or interests;
- issuance of project completion certificate; and
- major contractual variations.

For example, the SPV is not allowed to transfer any project agreement to another party without the lender's approval.

Security Agreement

This agreement contains the security package for the project to address the possibility of project failure. In lending, a security or collateral is an asset pledged by the borrower in case of default. The Agreement may include

- escrow of revenue;
- first charge on mortgage over project land and properties;
- charges on movable assets;

- sponsors' support, e.g. contingent funds, pledge of shares and SPV accounts, completion guarantee, and parent company guarantee;
- assignment of benefits from major project contracts, such as purchase and license payments, performance bond proceeds, PPP contract termination payments, and liquidated damages backed by a Letter of Credit;
- assignment of insurance proceeds and non-parent guarantees, such as a Partial Risk Guarantee provided by the World Bank to lenders against political risk; and
- loan covenants (see section on "Common Terms Agreement").

The escrow of revenue requires the SPV (borrower) to set up an escrow account with the lender. The latter will deduct the loan repayment and other payments before releasing the remaining sum to the SPV. The objective is to prevent revenue streams from disappearing before the repayment of loan.

Sponsors need to pledge their shares with the lender. The borrower retains ownership of the shares. However, upon default of the loan, the lender may take over the ownership. Further, if the pledged shares fall in value, the lender may demand additional funds.

A Letter of Credit (LC) is commonly used in trade to ensure payment to the seller upon delivery of the goods to the buyer. The bank provides the LC to the seller with the undertaking that it will pay if the buyer does not. The bank then seeks redress from the buyer.

Enforcing security may be an issue. For example, mortgage laws may be weak in developing countries, the concept of charges is not applicable in many countries, and grantors may object to the assignment of PPP contract benefits to lenders.

To secure insurance proceeds, lenders need to control the insurance package through the choice of insurer, nature and extent of cover, and assignment of benefits to lenders even if the policyholder breaches policy conditions.

Equity Support Agreement

In this agreement, sponsors agree to

- progressively provide a base equity of about 20 to 40%;
- provide project completion guarantee to lenders;

- provide contingent funds for cost overrun; and
- repay all debts if they abandon the project.

Common Terms Agreement

This agreement between lenders and the SPV deals with

- disbursement and payment;
- conditions precedent;
- representations and warranties;
- covenants; and
- default of the borrower.

The disbursement and payment section contains items related to draw-downs, repayment, grace period, fees, and prepayment. Lenders charge fees to cover their cost of administering the loan, as well as a commitment fee over the undisbursed balance of the loan.

Before disbursing the loans, lenders require that certain conditions must be met, such as the delivery of project contracts and agreements, security documents, legal opinion on validity and enforcement of these documents, initial deposit of base equity, and payment of lending fees.

The disbursement, such as to the contractor, requires certification by the SPV's auditor and lenders' technical adviser on project cost and progress. For project works in progress, a request for progress payment requires certification by the specialist consultant, cost consultant, and project manager.

The SPV is required to warrant basic facts on its creditworthiness. These representations may include its corporate status, non-default status, non-creation of additional security on project assets, fair financial statements and financial soundness, and environmental compliance.

The loan covenants that require compliance include the use of free cash flow, inspection and reporting, and other matters. On free cash flow, the SPV is required to maintain project accounts (see Project Accounts Agreement), ensure adequate cover ratios, and furnish an annual operating budget for the lenders' approval. The agreement may provide for higher repayment ("cash sweep") if the project has adequate free cash flow. Conversely, if the project falters, there may be a "Clawback clause" for investors to return dividends or other cash deficiency provisions to provide additional funds.

There are also negative covenants spelling out what the SPV must not do, such as not distributing profits without ensuring debt repayment and fully funded cash trap accounts, that is, sinking funds and debt service reserve.

The following items normally require lenders' approval:

- disposal of major assets;
- major investment;
- further borrowings;
- change of business;
- setting up of subsidiaries;
- transfer of shares; and
- execution of contracts other than approved project documents.

Lenders have the right to inspect project assets and records. The SPV must furnish audited financial statements and construction progress reports certified by the lenders' technical adviser.

The other matters include the payment of taxes, maintenance of insurance, and project compliance with regulatory requirements.

Finally, default triggers include non-payment of debt, failure to comply with equity commitment, failure to pay third parties, misrepresentation, insolvency, non-compliance with covenants, and government actions such as appropriation. For remedies, there is a short cure period beyond which lenders may stop the loan and demand payment, step in to run the project, or sell assets to recover the loan.

Accounts Agreement

This agreement provides for the setting up of several accounts in order of priority on the use of revenue from an escrow account:

- operations;
- major maintenance (sinking fund);
- debt service payment;
- debt service reserve;

- debt service payment on subordinated debt, if any;
- debt service reserve on subordinated debt, if any;
- voluntary prepayment, if any;
- capital investment; and
- restricted payments (e.g. dividends).

The priority is important because if there is insufficient revenue, then it is unlikely that dividends will be payable. The above priority payment is also called the waterfall cash flow model. Each "bucket" must be filled before money flows down the priority ladder.

The accounts may be handled by a trustee who is usually a senior lender in the syndicate, that is, the accounts bank. From the lenders' perspective, letting the SPV handle the accounts creates a lot administrative work and is less secure for debt repayment.

Inter-Creditor Agreement

This is an agreement among lenders in the syndicate on how to conduct their business and prevent unilateral actions by a single lender. The main provisions include each lender's share of the syndicated loan, order of drawdowns, allocation of debt service repayments, default, approval of SPV's operating budget, and how to deal with changes or requests from the SPV. For example, the SPV may wish to issue a bond to refinance the project, and this requires lenders' approval.

Finalization of project documents

Lenders may require changes to specific clauses in project documents to mitigate certain risks. Clearly, these changes will need to be negotiated, and, once agreed, the documents will be finalized.

Hedging arrangements

Recall that hedging instruments include interest rate swaps, currency swaps, futures contracts, and forward contracts. Lenders will want to ensure that appropriate hedging instruments are in place.

Financial close

Once the conditions precedent to a loan have been satisfied, lenders and sponsors will reach a financial deal. This ensures that everything is in order, including the PPP contract, permits and approvals, project contracts and agreements, and insurances.

Refinancing

Upon completion of construction, sponsors may seek to refinance the loan at a lower rate of interest because the construction risk has been eliminated. One way is to renegotiate the loan with the existing lender.

Another possibility is for the SPV to issue bonds, which incurs issuance costs but, more importantly, is less flexible when it comes to renegotiating with many investors such as if the project runs into serious trouble.

A third refinancing approach is for the SPV to secure a new loan from long-term investors such as insurers, infrastructure funds, and pension funds.

Securitization

Lenders may not want to hold on to project loans because of the long duration, which may extend up to 40 years. There is a clear mismatch between long-term loans and short-term deposits. In some countries, such as the United States, the lender provides a land and construction loan, and upon completion of construction, a permanent lender, such as an insurer, takes over the loan. However, such loans are uncommon; it is more usual for a lender to provide long-term rather than take-out financing in other countries.

Loan securitization offers a solution to overcome the mismatch of loans and deposits. The originating lender sells the loans to a special purpose entity (SPE) set up by a sponsor. The sponsor of the SPE is a financial institution. It is not the sponsor of the SPV. They are different parties. The securitization SPE, not to be confused with the project SPV, then issues securities to investors to obtain the funds to pay the originators. The originators may then use the funds to finance new projects or for some other purposes. In addition to improving liquidity, the originator no longer assumes project risks because it no longer owns the loans to earn interest income.

The SPV (borrower) will continue with periodic repayments of the loan to the originator. In turn, the originator, for a small fee, will forward these repayments to a trustee, which is another financial institution. The trustee then uses the proceeds to pay the investors.

The SPE pools (buys) project loans from different types of projects from originators. These loans are then categorized into tranches (e.g. Class A, B, and C) with varying risk profiles and priorities of payments and, hence, different interest rates. These rates are normally quoted in terms of the spread over a benchmark rate, such as the Singapore Interbank Offered Rate (SIBOR). Investors select the tranche(s) that suit(s) their risk appetite. The bulk of the securities will be Class A grade, which is less risky. Other possible credit enhancements to attract investors include

- insurance against default by borrowers on investment grade securities;
- ratings on securities;
- repurchase agreements with originators to buy back affected loans;
- warranties by originators against certain project risks;
- interest rate swaps to convert floating to fixed rates;
- participation by originators, especially in purchase of Class C securities;
- sale of loans by originators at a discount;
- currency swaps to mitigate currency risks if the original loans are in different currencies; and
- making securities tradable in the secondary market.

Investors will bear other risks, such as delays in payments and non-performance by the parties, such as the originator, trustee, and sponsor of SPE. Normally, the SPE appoints a servicer to do the administration and processing. The servicer may not perform.

Investors will have to do their homework on the above credit enhancements, project quality, size of the pool of projects, whether construction has been completed at time of securitization, and so on. As the subprime mortgage crisis has shown, securitization can go badly wrong (Mullo and Padilla, 2009; Schultz and Fabozzi, 2016).

References

Harris, D. (1974) Credit rationing at commercial banks: Some empirical evidence. *Journal of Money, Credit and Banking*, **6**(2), 227–40.

Khan, F. and Parra, R. (2003) *Financing large projects*. New York: Pearson.

Mullo, P. and Padilla, M. (2009) *Chain of blame: How Wall Street caused the mortgage and credit crisis*. New York: Wiley.

Ryan, R. (2009) The role of lead counsel in syndicated lending transactions. *The Business Lawyer*, **64**(3), 783–800.

Schultz, G. and Fabozzi, F. (2016) *Investing in mortgage-backed and asset-backed securities*. New York: Wiley.

Chapter 10

Pre-construction activities

Establishment of the special purpose vehicle

The next step after the financial close is for sponsors to establish the special purpose vehicle (SPV) as an entity to sign the public–private partnership (PPP) contract with the grantor. The parent company is not directly involved in running the project because of liability and other risks in different countries, such as local ownership requirements and restrictions on property transactions. Lenders also require sponsors to set up an SPV to ring-fence the project revenues and cash flows as security for the loan.

Types of businesses

Sponsors have different ways to structure their working relations (Table 10.1), and the type of business, ownership, liability, and taxation will differ depending on tax jurisdictions.

Table 10.1   Types of investment vehicles.

Type of business	Owner	Liability	Taxation
Incorporated			
Corporation	Shareholders	Limited	Profit and dividend
LLC	Shareholders	Limited	Income tax
Unincorporated			
Sole proprietorship	Sole proprietor	Unlimited	Pass through
Partnership	Partners	Unlimited	Pass through

In an unincorporated business entity, the business owners have unlimited liability; that is, they are responsible for its debt. For example, in a general partnership (GP), all partners are liable for its debt and may have to sell their personal assets to repay it. In a limited partnership (LP), at least one partner has unlimited liability.

A common mode for project financing is the limited liability company (LLC) because of the limited liability and avoidance of corporate double taxation on profit and dividend. In an LLC, the profits are not taxed at the source and are distributed or "passed through" to shareholders to file their own taxes.

The incorporation is usually in a low-tax country such as the British Virgin Islands (BVI), where it is easier to manage tax and other matters such as the avoidance of double taxation, minimal disclosure and reporting requirements, and flexibility in profit distribution. A project company pays tax where it operates, not where it is incorporated. Hence, a BVI-incorporated company does not avoid paying tax in the country of operation. If a parent company has projects in different countries, it is likely to set up a passive holding company in a low-tax country to consolidate the multiple sources of income.

Project governance structure

At the organization level, corporate governance refers to the rules and processes that direct and control a company. Similarly, the project governance structure spells out who has the authority to make decisions to direct and control the project.

Typically, there are two levels of decision-making within the SPV. Management will make the strategic decisions and serve as the approving authority. It issues a project charter to appoint the project team. The team works in the project office and is responsible for the day-to-day execution of the project. It comprises a senior and experienced project manager, in-house members, and consultants. The composition depends on tasks, experience, and expertise. If an organization has many projects, it is likely to set up an enabling project management office (PMO) to develop, coordinate, and standardize procedures (Taylor, 2016).

Project Brief

The sponsor or Project Director will prepare a Project Brief. It often contains the following information:

- project title;
- sponsor;
- background;
- goals and objectives;
- scope, design, and facility performance requirements;
- project performance criteria;
- limits to authority of the project manager;
- site information;
- project governance structure;
- procurement strategy;
- draft contracts and agreements;
- reporting and monitoring;
- key stakeholders;
- regulatory requirements;
- budget, preliminary schedule, and milestones;
- expected quality;
- opportunities and constraints; and
- major risks and mitigation.

The project success performance criteria often include the technical criteria of time, cost, quality, and safety expectations. In addition, a project must satisfy the business criteria of stakeholder satisfaction and alignment with the business case.

Appointment of the project manager

The next task is to appoint the project manager if this has not been done earlier. The project manager may be from the SPV or is an external consultant. The in-house project manager may be seconded from the main sponsor's parent company. The Project Director uses the Project Brief to brief the project manager.

Requirements

After studying the Project Brief, the project manager will gather additional requirements from stakeholders.

Internal stakeholders are from within the organization (SPV), such as facilities managers, and external stakeholders include regulators, affected businesses and residents, and potential users.

Scope

The next step is to develop a preliminary Work Breakdown Structure (WBS) on the main tasks ahead. Once these tasks are understood, the project manager will recommend that a suitable project team be appointed. The composition of the team will depend on the expertise required.

For the SPV, a high-level WBS is used to determine the main components of the project. An example for a school building is given in Fig. 10.1.

1 School Project
2.1 Preconstruction
2.1.1 Design
2.1.2 Construction documents
2.1.3 Tender
2.1.4 Contract award
2.1.5 Mobilization
2.2 Construction
2.2.1 Main building
2.2.2 Multi-purpose hall
2.2.3 Gymnasium
2.2.4 Pool
2.2.5 Track and field
2.2.6 Carpark
2.2.7 Learning garden
2.2.7 Road
2.2.8 Landscaping
2.3 Close out
2.3.1 Commissioning
2.3.2 Green Mark certification
2.3.3 Substantial completion
2.3.4 Rectify punch list
2.3.5 Training of facilities management team
2.3.6 Final account
2.4 Post-construction
2.4.1 Warranties
2.4.2 Defects follow-up
2.4.3 Occupation

Fig. 10.1   Example of a high-level WBS.

Procurement strategy

If the sponsors have not decided on the procurement strategy, the project manager may recommend one. This book uses the traditional design-bid-build procurement strategy as an example. This strategy provides the SPV with greater control over design and cost, and is preferred by lenders (see Chapter 7).

Appointment of the project team

Once the procurement strategy is finalized, the SPV will proceed to appoint the designers. For building projects, the designers are architects and engineers. The typical arrangement is to hire external designers for a negotiated fee based on a percentage of construction cost or through competitive bidding.

Project schedule

The project team develops the project schedule using a bar chart (Table 10.2). It lists the main activities to undertake up to the start of the operation and maintenance phase.

Table 10.2   Project schedule.

	Period			
	1	2	3	...
Site survey and engineering studies				
Site visits				
Visit similar facilities				
Programming				
Schematic design				
Design development				
Construction documents				
Final review				
Permits and approvals				

(*Continued*)

Table 10.2 (*Continued*)

	Period			
	1	2	3	...
Tender				
Award of contract				
Mobilization				
Construction				
Purchase of equipment				
Commissioning				
Occupation				
Staff training				
Start of operation				

Site survey, analysis, and engineering studies

A site survey is necessary to establish legal boundaries, develop contour maps, and subsequently peg the building layouts. The site analysis is similar to that covered in Chapter 4, except that it is the sponsor rather than the grantor that carries out the analysis.

The engineering studies will include a geotechnical report and possibly reports on environmental, traffic, town planning, and other community concerns. The grantor may have conducted some of these studies and shared them with the SPV, usually on a "for information only" basis without liability. This means that, if necessary, the SPV must do its own due diligence by conducting an independent site study.

Programming

Programming refers to the process of gathering project requirements. The project team extracts the basic requirements from the Project Brief. This is supplemented by the additional high-level requirements gathered by the project manager.

The project team may visit similar facilities to gather more ideas. Finally, it gathers additional requirements from different stakeholders.

To guide this process, the team breaks down the project manager's high-level WBS further and collects requirements for each component (Fig. 10.2).

2.2.1 Main building
2.2.1.1 Substructure
2.2.1.2 Superstructure
2.2.1.3 Building envelope
2.2.1.4 Building systems
2.2.1.5 Interior finishes

2.2.1.4 Building systems
2.2.1.4.1 Mechanical system
2.2.1.4.2 Plumbing system
2.2.1.4.3 Electrical system
2.2.1.4.4 Security system
2.2.1.4.5 IT system
2.2.1.4.6 Fire-fighting system
2.2.1.4.7 Building management system
2.2.1.4.8 Special systems

Fig. 10.2   Example of a lower-level WBS.

Some buildings require special systems, such as an acoustic system for a concert hall.

Design Brief

Once the requirements are collated and finalized (approved), the project team develops the Design Brief to guide the design. For example, for a building, the Design Brief consists of narratives and information on the following items:

- project;
- design concept;
- target green rating and simulation requirements;
- life-cycle considerations;
- opportunities and constraints;
- building orientation;
- integration with adjacent uses;
- privacy;
- visuals and lines of sight;

- major systems;
- special systems;
- major materials and colors;
- space program;
- use of external spaces;
- landscaping;
- regulatory approvals;
- Building Information Modeling (BIM) requirements;
- equipment and furnishings, including special equipment; and
- design review process.

We assume that the grantor has fixed the design concept during the PPP tender when requesting for price and design proposals.

The space program divides the building into different spaces, each with its data sheet of requirements and adjacencies to locate functionally related rooms near each other. For instance, a library should not be located near a noisy canteen.

Design management

The project team manages the design process by

- assigning design responsibilities for each design package;
- establishing the number of design options to be considered;
- developing progressive cost and schedule estimates;
- managing the design review and approval process;
- performing value engineering and constructability review if necessary;
- managing the regulatory planning, transport, and design approvals; and
- managing the design information.

Normally, a senior lead designer, such as an architect, has overall design responsibility. She uses a design work breakdown structure (DWBS) to assign the design of different components to other designers. For process plants, the lead designer is an engineer.

In general, sponsors expect designers to generate about three options before approving and finalizing the conceptual or schematic design. It is assumed that the grantor has fixed the land-use plan and schematic design during the tender by selecting the winning bidder. The next stage is design development, followed by the development of contract documents.

Schematic design

Design begins with a land use plan for the site. For larger projects, there may be a master land use plan.

Each site is subject to land use regulations, such as

- a building line from the center of a road to the front of the building;
- setback distances from the building to the site boundaries;
- ground coverage ratio, or ground floor area/site area;
- plot ratio, or gross floor area/site area;
- height restrictions;
- parking requirements; and
- development charge or fee for the change of use or intensity of land use.

The schematic design shows the concept of the development. For a building, it includes elevations (e.g. north elevation), floor plans, stack plans, and sections for the owner's approval to proceed to design development. A stack plan shows the different space uses of each floor of a building.

Design development

In design development, the schematic design is further developed, including the detailing of architectural forms, structures, building systems, and floor plans.

This stage includes progressive cost and schedule estimates as well as the conduct of design reviews, value engineering, and constructability review. It also includes seeking regulatory approvals.

Progressive cost and schedule estimates

As the design develops, the project team will progressively estimate the project cost and schedule with increasing precision. This ensures that the project is designed to budget and can be built within schedule. Table 10.3 shows an example using elemental costing for a building.

Table 10.3   Elemental cost estimate for a building.

	Estimated cost	Cost/m^2	Percentage
General requirements			
Site work			
Concrete			
Masonry			
Metals			
Woods and plastics			
Thermal and moisture			
Doors and windows			
Finishes			
Specialties			
Equipment			
Furnishings			
Special construction			
Conveying systems			
Mechanical			
Electrical			
Total direct cost			
Material tax			
Labor tax			
Contingency			
Insurance			
Bond			
Profit @10%			
Estimated tender price			

The estimated schedule will depend on the complexity of the project. For the SPV, the schedule estimate for a large infrastructure project will only be in the order of months, such as 48 months. The contractor will provide the SPV with a more accurate schedule after the award of the construction contract.

Design reviews

During design development, there are often two review and approval stages, when the design is at 30% and 90% of completion, respectively, to ensure that the design

- adheres to design criteria;
- is within scope;
- meets functional and operational objectives;
- complies with the code;
- is integrated and has compatible interfaces;
- considers safety;
- does not contain conflicts, errors, and omissions; and
- is on schedule.

The design review is also an opportunity for the SPV (sponsors) to obtain feedback from designers.

Value engineering

The design team may carry out value engineering (VE) at the 30% and 60% design stages to provide a leaner design (Younker, 2003; Huthwaite, 2004). This may be accomplished by

- removing unnecessary expenditure;
- simplifying methods and procedures;
- removing redundant items;
- considering the use of alternative materials and standards where applicable;
- reducing excess inventory;
- reducing unnecessary transport;

- minimizing defects and rework;
- relooking at sourcing;
- re-examining standards;
- reducing environmental waste; and
- reducing energy requirements.

VE started as value analysis (VA) and was subsequently renamed as value analysis and value engineering (VAVE), especially in product development. Nowadays, the terms VE and value methodology (VM) are more common.

The process begins by selecting the main elements, identifying their functions, and generating alternative solutions. These solutions are evaluated, and the approved suggestions are implemented. VE is not a cost-cutting exercise to compromise on quality. The functions must be preserved, possibly with improved performance and simpler designs.

Constructability review

Depending on the scale and complexity of projects, there may be a constructability review of the draft bid documents near the end of the design stage to

- ensure conformance with codes;
- remove design conflicts;
- resolve interface issues;
- determine if the schedule is realistic;
- ensure that drawings and documents are coordinated;
- remove errors;
- check for omissions; and
- ensure safe construction, maintenance, and operation of the facility.

An improved design will attract better-informed bids from contractors and speed up the construction process. The documents to review should also include soil, environmental assessment, product, and test reports to ensure that vital information is not overlooked and there is consistency of information. A constructability review should not be used to redesign the project.

The review team may be a single person or, more commonly, another internal project team with fresh pairs of eyes. If the internal team is busy with another project, an external team may be appointed at additional cost.

The sequence of review follows the construction process; that is, from the substructure to the superstructure, building envelope, architectural interiors (e.g. space allocations), mechanical, electrical and plumbing (MEP) systems, and external site work. Where design conflicts arise, there is a normative sequence of design priority. In general, structural elements take precedence over architectural and MEP elements. The logic is that elements that are more difficult or costly to alter should take precedence.

BIM is often used to replace the traditional 2D drawings for a con-structability review. However, a BIM model may lack certain details, such as structural connections, roofing details, water-proofing, and electrical or piping branches.

Regulatory design approvals

The required regulatory design approvals at different stages of the design process often relate to town planning, building, fire, utilities, health, trans-port, pollution, and environmental matters. In many cases, designers sub-mit these plans for approval on behalf of the owners.

The trend is towards simpler and faster regulatory approvals, such as one-stop approval centers, instead of having to consult multiple agencies.

Managing design information

There are many design drawings and there is a need to develop a system to manage and share the information among relevant parties. The general requirements are to categorize different types of drawings using an appro-priate numbering system (e.g. "Axxx" for architectural drawings), keep track of different versions and amendments, and implement a system for dissemination. After design, the information is required for construction, operation, and maintenance.

A paper-based system will be costly because the information is diffi-cult to organize, update, and retrieve. In the 1980s, some of the paperwork was converted into electronic files, but the problems remain. The current preference is to use BIM. Some countries mandate the submission of plans using BIM for approval. BIM has other uses, such as to develop the schedule (4D BIM), progressive cost estimates and control (5D BIM), analysis of building performance such as energy simulation (6D BIM), and operation and maintenance (7D BIM).

Before implementing BIM, the parties need to sort out many commercial, legal, and technical issues (Hardin and McCool, 2015). The commercial and legal concerns include roles and responsibilities, rights to information, liability, insurance, training and capacity, collaboration process, and security of information. The technical issues include object classification, location (coordinates), attribute data, construction tolerances, volume computation, drawing templates, naming conventions, annotations, software versions, data exchange formats, and control over the common data environment.

Commissioning plan

The project team prepares the commissioning plan to test the major systems and equipment. For example, for a building, the systems to be tested include lifts, escalators, air-conditioning, lighting, fire protection, structural system, security, IT systems, and so on.

The plan consists of the following (Grondzik, 2009):

- systems to be commissioned and performance criteria;
- system integration tests;
- system tests;
- whether the commissioning is done by the designers or an external party acceptable to the SPV's project team and contractor;
- commissioning budget;
- risk assessment;
- special resources required;
- commissioning schedule; and
- commissioning reports.

System integration tests ensure that different components work properly. For example, in software development, data is often passed from one system (e.g. leasing) to another (e.g. facilities management). The test will require the use of test scripts and test data to check its proper transfer and the logic of the process.

System testing concerns the workings of the entire system. Although the components may integrate well, the entire system may still fail.

Development of contract documents

As the design proceeds, it will be necessary to capture the design intent as part of the contract documents. These contract documents include the specifications and drawings. We will briefly discuss the documents in the next section.

Tender documents

The construction tender documents consist of the following:

- Invitation to Tender;
- Instructions to Bidders;
- Bid Form;
- Form of Agreement;
- Form of Bond (performance bond);
- General Conditions of Contract;
- Special Conditions of Contract;
- Drawings and specifications;
- Schedule of Basic Rates;
- Bills of Quantities (if any); and
- Addenda.

The General Conditions of Contract contain common contract clauses. A project may have special clauses, and they are found in the Special Conditions of Contract. Each specification has three elements, namely,

- the product;
- the standards or certifications required; and
- the standard of workmanship, testing, training, and other requirements.

Bidders will fill in the Schedule of Basic Rates as part of the tender for subsequent use in pricing variations orders. The project team is likely to renegotiate the rates before awarding the construction contract. If the Bills of Quantities are provided, contractors will use them to price their tenders. Finally, the Addenda contain any last-minute changes to the tender documents.

Provisional and prime cost sums

If the design of a component is not complete at the time of tender, contractors will be asked to put a provisional sum for that item in the bid.

A sponsor may nominate a subcontractor for certain materials or specialized work. The contractor will put the cost of this arrangement as a prime cost sum in the bid. He will assist the nominated subcontractor on site, such as by coordinating the work, the lifting of goods, and the supply of utilities. Hence, the contractor is entitled to mark-up and attendance. For instance, if the nominated subcontractor's contract value is $x, the contractor may be entitled to 10% of $x. The percentage of mark-up will depend on the contract value.

Tender process

The tender process for the SPV to select the contractor under the traditional design-bid-build contract is similar to the grantor's tender process. It begins with pre-qualification, followed by an invitation to tender, pre-bid conference, receipt of bids, evaluation, negotiation, and awarding of contract.

Contractor's decision to bid

A contractor will decide whether to bid for the project by examining the workload, labor conditions, project prestige, reputation of sponsors and designers, number of bidders, risk allocation, type of contract, contractual terms, safety, possible site problems, performance bond capacity, and the quality of contract documents.

Contractor's bid estimate

If the contractor decides to bid, he will

- review the tender documents;
- visit the site;
- attend the pre-bid meeting;
- seek owner's clarifications; and
- develop the bid.

The bid is estimated from the contractor's WBS. This is similar to the owner's WBS but in greater detail. The lowest level of the breakdown is the work package (WP). For each WP, the contractor will have to decide whether he will do the work or subcontract it. If it is subcontracted, the contractor will seek subcontractors' quotations and use the lowest or best-value quote in the bid. These WPs are then rolled up or aggregated to form the bid. In Table 10.4, the estimate is for a building using an industry-wide specification format (Construction Specifications Institute, 2011). There are many similar formats that vary across countries, and they are periodically updated, usually with more divisions or subdivisions.

The office overhead cost is included in the general requirements division of work. Further, capital cost is included in the equipment division.

Table 10.4   Contractor's bid estimate for a building.

Division	Material	Labor	Subcontract	Total
General requirements				
Site work				
Concrete				
Masonry				
Metals				
Woods and plastics				
Thermal and moisture				
Doors and windows				
Finishes				
Specialties				
Equipment				
Furnishings				
Special construction				

(*Continued*)

Table 10.4 (*Continued*)

Division	Material	Labor	Subcontract	Total
Conveying systems				
Mechanical				
Electrical				
Total direct cost				
Material tax				
Labor tax				
Contingency @5%				
Insurance				
Bond				
Profit @10%				
Estimated tender price				

From the direct cost, we add taxes, contingency, insurance premium, performance bond premium, and mark-up for profit to obtain the estimated tender price.

For non-building works such as a hydropower dam, the WBS will include river diversion, dam, spillway, penstock, turbine, generator, transformers, hydraulic hoists, transmission lines, and so on. For roads and bridges, there are also industry-wide specifications formats, such as those provided by the US Department of Transportation (2017).

Award of construction contract

The process of awarding the construction contract by the SPV to the contractor is similar to the grantor's award of the PPP contract to the SPV. The SPV will evaluate the bids from contractors based on price or best value. It will then sign a Letter of Intent with the winning bidder. The construction contract will be signed after the contractor has submitted the required documents as stated in the contract. Typically, they include the project insurances (see Chapter 7), performance bond, list of key subcontractors and suppliers, and baseline plans (see Chapter 11). The performance bond is about 10% of the contract sum. The surety or bonding company will pay the sponsor if the contractor does not perform and then seeks to recover the sum from the contractor. The bond normally extends until the end of the defects liability period, which is usually one year after the completion of construction. The performance bond is callable depending on whether the sponsor needs to prove the contractor's breach of contract. If proof is required, it is a conditional bond; otherwise, it is an on-demand bond. Obviously, sponsors prefer on-demand bonds.

Finally, the SPV will issue a Notice to Proceed to the contractor to start the contract clock.

References

Construction Specifications Institute (2011) *The CSI construction specifications practice guide*. New York: Wiley.

Grondzik, W. (2009) *Principles of building commissioning*. New York: Wiley.

Hardin, B. and McCool, D. (2015) *BIM and construction management*. New York: Wiley.

Huthwaite, B. (2004) *The lean design solution*. Michigan: Institute for Lean Innovation.

Taylor, P. (2016) *Leading successful PMOs*. London: Routledge.

US Department of Transportation (2017) *Standard specifications for construction of roads and bridges on federal highway projects*. Washington DC: FHA.

Younker, D. (2003) *Value engineering*. London: CRC Press.

CHAPTER 11

Mobilization

Submittals

The special purpose vehicle (SPV), as the project owner, will require the contractor to submit the following during mobilization:

- performance bond (see Chapter 10);
- project insurances (see Chapter 7);
- schedule of values (SoV); and
- project baseline plans.

The last two items are discussed below.

Schedule of values

The contractor prepares and submits an SoV to the SPV so that the latter and lender can forecast the cash flows for each construction progress payment (e.g. quarterly). The SoV is obtained from the contractor's bid estimate, which is now converted to contract value (Table 11.1). For example, the first item on "General requirements" may originally cost $10 m, excluding the indirect cost. To convert it to value, it is multiplied by 130/100 to give $13 m. This value is then distributed according to the projected progress of work on a monthly or quarterly basis.

In submitting the SoV, the contractor will not include the indirect cost, which is private information. He may *front-load* certain items so that he is paid earlier, and this is something the owner needs to scrutinize. For instance, he may put in $15 m instead of $13 m for the first item.

Table 11.1   Schedule of values.

Item	Division	Value ($m)	Q1	Q2	...
			Timeline		
1	General requirements	13	8	2	
2	Site work				
...	...				
16	Electrical				
Total direct cost	100				
Indirect cost	30				
Contract value	130	130			

Baseline plans

The baseline plans, also called the project management plan, is a series of plans on how the contractor will manage the project. They are plans for managing the following:

- construction schedule;
- resources;
- cost;
- quality;
- safety and health;
- scope;
- environment;
- stakeholders;
- commissioning;
- site organization and staffing;
- risks;
- subcontractors and suppliers;
- inspection;
- documents; and
- claims and dispute.

The contract will stipulate the plans to submit to the owner. The latter may provide feedback to adjust the plans. For instance, the owner may have some concerns about the safety and health plan. Some plans are of little interest to the owner, such as how the contractor manages its subcontractors and suppliers.

The contractor's site organization and staff plan is discussed in this chapter. The rest of the plans will be discussed in Chapter 12, when they are implemented.

Site organization and staffing

In organizing the site, the contractor needs to consider the site constraints that will affect the following:

- connection to utilities;
- jobsite safety;
- signage;
- security;
- ingress and egress routes;
- movement and storage of materials and components;
- productivity of workers;
- use of equipment;
- preparatory work area;
- location of the site office, rest area or canteen, toilets, workers' quarters, and temporary facilities;
- external pedestrian and vehicular traffic; and
- environmental control measures.

The site and security offices are located near the entrance for easier control of visitors. Materials should be protected and stored near work and installation areas for easy access and movement.

The environmental control measures include solid waste disposal, prevention of flooding and soil erosion, proper discharge of water, pest control, proper handling of hazardous materials, care of vegetation and conservation of heritage, minimizing noise and dust, and proper sanitation. These measures must satisfy regulatory requirements and be subject to audit.

Cranes require overhead rights such as over property, roads, and power lines to move materials. However, third parties may not grant these rights, thereby restricting movements to within the site boundary. Because of site constraints, the use of "just-in-time" materials delivery (Ohno, 1988) requires the coordination of construction, crane, and delivery schedules.

The staffing arrangement is given in Table 11.2. There are many variations depending on the complexity and size of the project.

Table 11.2   Contractor's staffing arrangement.

Role	Responsibilities
Head Office	Contract administration, cost estimating
Project manager	Overall-in-charge, liaises with designers, and reports to Head Office
Superintendent	Site activities, including field engineer, supervisors, and quality inspector
Assistant superintendent	Subcontractors and suppliers
Office manager	Documents, accounts, purchasing, and so on
Project engineer	Field engineering and project control
BIM manager	BIM coordination
Health and safety officer	Site health and safety
QA/QC inspector	Quality
General foreman	Foremen or supervisors of different trades

Kick-off meeting

The kick-off meeting is the first meeting for the sponsor's project team to brief the contractor's project team. The key subcontractors may also attend the kick-off meeting if the subcontracts have been awarded. The agenda includes

- brief project information covering the rationale for the project, sponsor(s), and goals and objectives;
- key stakeholders;
- partnering agreement, if any;
- project scope;

- broad schedule and milestones;
- success criteria;
- constraints;
- roles of team members and communication lines; and
- procedures such as change orders, commissioning, and progress payment.

A "partnering agreement" among key stakeholders spells out the project's goals and objectives, the commitment to work together in the spirit of trust and openness, and timely resolution of disputes (Godfrey, 1996). When the commercial stakes are high, the parties are more likely to regulate their relations by contract if they do not have equal bargaining power. While top management may sign a partnering agreement, perceptions at the project level may be different.

Award of subcontracts

Depending on the sequence of construction, the contractor will proceed to award subcontracts according to the construction sequence. Generally, there will be flow-through clauses from the main contract to the subcontracts.

Procurement

The contractor will need to procure equipment, materials, and workers. Long-lead items such as specialty equipment, custom-made components or systems, and imported items should be procured early to avoid supply disruptions and delays.

References

Godfrey, K. (2006) *Partnering in design and construction.* New York: McGraw-Hill.
Ohno, T. (1988) *Toyota production system.* New York: Productivity Press.

CHAPTER 12

Construction

Project monitoring and control

During the construction phase, the contractor implements the project baseline plans and makes suitable adjustments as the construction progresses. These plans and their implementation are discussed below. In this chapter, the terms project owner, sponsor, and special purpose vehicle (SPV) are used interchangeably.

Construction schedule

Recall from Chapter 7 that the SPV has a broad project schedule with milestones. The contractor submits a more detailed schedule for the SPV to monitor construction progress.

The contractor's schedule is developed by creating a precedence diagram (Table 12.1) from the work breakdown structure (WBS). In practice, there will be many activities (or work packages), and we use a simple example to illustrate how to identify the critical path. It is important to identify the activities, predecessors, and durations accurately from experience.

Table 12.1   Precedence diagram.

Activity	Predecessor	Duration (weeks)
A	—	3
B	A	4
C	A	2
D	B, C	5
E	C	1
F	D, E	2

3	B	7
	4	

7	D	12
	5	

0	A	3
	3	

12	F	14
	2	

3	C	5
	2	

5	E	6
	1	

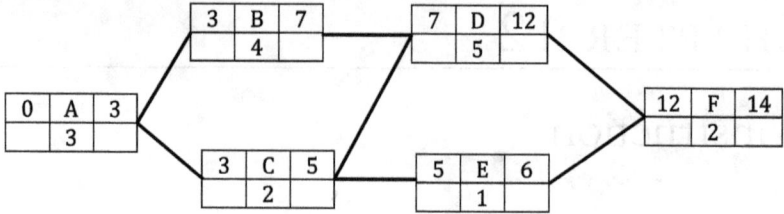

Fig. 12.1   Forward pass.

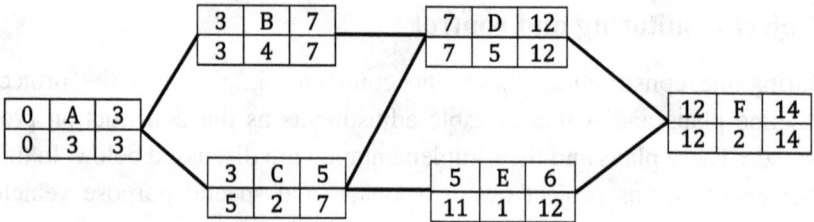

3	B	7
3	4	7

7	D	12
7	5	12

0	A	3
0	3	3

12	F	14
12	2	14

3	C	5
5	2	7

5	E	6
11	1	12

Fig. 12.2   Backward pass.

The next step is to use the arrow diagrams (Figs. 12.1 and 12.2) to determine the *critical path*. The convention is as follows:

ES	Activity	EF
LS	Duration	LF

Here, ES is early start, LS is late start, EF is early finish, and LF is late finish. The forward pass begins with activity A and continues until activity F. Observe that F has two predecessors (D and E) and can only start on Week 12.

The backward pass begins with activity F, starting with LF = 14 weeks, the project duration. We then work backwards to compute LF and LS for activities D and E. For activity C, LF is the minimum of (LS D, LS E) = min (7, 11) = 7. Similarly, for activity A, LF = min (LS B, LS C) = min (3, 5) = 3. The critical path consists of activities with EF = LF, that is, A–B–D–F. Any delay in a critical path activity will delay the entire project. The critical path may be different if the durations change; that is, another "near" critical path may become critical. For activities not on the critical path, LF – EF = LS – ES is the slack or float. For example, activity C can start as early as Week 3 or as late as Week 5.

Table 12.2   Project schedule.

Task ID	Description	Duration	1	2	3	4	5	6	7	8	9	10	11	12	13	14
																Week
1	A	3	x	x	x											
2	B	4				x	x	x	x	x						
3	C	2				x	x	x	f	f						
4	D	5								x	x	x	x	x	x	
5	E	1						x	x	f	f	f	f	f	f	
6	F	2												x	x	x

The final step in preparing the schedule is to develop a bar or Gantt chart using the early start timings (Table 12.2). A boldface "**x**" indicates critical path activities. Activities C and E are not on a critical path and have floats ("f").

After receiving the schedule, the owner may have some concerns over the activities, sequencing, and durations. For example, some activities may be missing, the durations may not be realistic, or the sequencing may be improved. Both parties will negotiate and adjust the plan.

In particular, the owner prefers not to have a schedule with late starts because it risks delaying the project. This raises the issue of ownership of the float; that is, does the owner or contractor own it? In other words, can the owner ask the contractor to adjust the float to avoid a delay?

This dispute may be avoided if the contract specifies who owns the float. If it is silent, one view is that the contractor owns the float. As long as the contractor delivers the project on time, the owner has no right to interfere as to how the contractor allocates the resources to finish the project. In this case, the contractor has planned a late-start schedule, possibly because of resource constraints.

The contrary view is that the owner pays for the project and, therefore, owns the float. Hence, the owner can instruct the contractor to develop a less risky schedule.

A third view is that the project owns the float; that is, the float is allocated on a first-come-first serve basis. If the *owner* delays the project, he may use any available float, and the contractor cannot claim for an extension of time. On the other hand, if the *contractor* delays the project, she may use any available float to speed up the project and avoid a delay penalty. This view is more common (Rubin *et al.*, 1999), but it becomes complicated if *both* parties are responsible for the delay.

The above discussion is based on traditional critical path scheduling. In the critical chain approach (Goldratt, 1999), the duration for each critical activity (A, B, D, and F) is shortened, resulting in a large "project buffer" between the end of the last critical activity (F) and the project deadline. The other key concept is the focus on resource constraints, which is similar to the traditional approach and discussed in the next section. The priority of resource allocation should be to critical activities.

A	B	D	F	Project buffer

The debatable assumption is that people are conservative and build in buffers for critical activities.

Finally, the contractor should track and document why a particular activity is delayed. The construction period may stretch several years, and it is unwise to leave important matters to memory. Documentation helps if disputes about the activity or the entire project arise in future.

Resource management

For each activity, the team estimates the key equipment (e.g. crane), labor, materials, and subcontractors required (Table 12.3).

In each category, sub-categories may be created using additional rows. For example, sub-categories for cranes and machinery may be created for "equipment," and each cell will show the number of each equipment required for the week. The numbers for labor refer to key personnel and that for materials are for illustration only in appropriate units.

In this example, the contractor has subcontracted activity C to subcontractor Y and activity E to subcontractor Z. However, subcontractor Y is busy with another project and cannot start in Week 3 as originally planned. Hence, activity C will start in Week 4 and end in Week 6, giving only one week of float. This process of rescheduling activities because of resource constraints is called *resource leveling*.

If progress is slow, it is possible to shorten the durations of some critical activities by *crashing*; that is, by providing additional resources to

Table 12.3   Project schedule and resources.

Task ID	Description	Duration	1	2	3	4	5	6	7	8	9	10	11	12	13	14
1	A	3	x	x	x											
2	B	4		x	x	x	x	x								
3	C	2				x	x	x	f							
4	D	5							x	x	x	x	x	x		
5	E	1					x	x	f	f	f	f	f	f		
6	F	2												x	x	x
	Equipment		2	2	3	3										
	Labor		5	5	6	6	5	5	4	4	4	4	3	3	3	3
	Materials						4	3	3	3	3	3	2	2		
	Subcon Y					x	x	x								
	Subcon Z						x	x								

accelerate work. The activities to crash will also depend on cost implications. Since materials are generally not substitutable, the cost (C) is given by

$$C = wL + rK$$

where w is the wage rate, L is the labor input, r is the capital rental rate, and K is the capital input. However, it may not be easy to substitute capital for labor (Tan, 2010). Construction work is often done in teams, and the capital–labor ratio is relatively fixed. For example, the ratio of a crane to crane operator is 1:1. It does not make sense to replace a crane with a team of manual workers for lifting.

The contractor will also examine non-critical activities and check if it is possible to use fewer resources and lengthen the durations. This will reduce the construction cost without delaying the project.

Apart from resource leveling and allocation of labor, the contractor needs to manage materials. These materials need to be sourced, procured, transported, stored, and protected from spoilage and theft.

Many constructors hire a mix of foreign workers, unionized workers, and non-unionized local labor. They need to provide site supervision, possibly accommodation, and safe working conditions. Workers should be treated fairly and incentivized to improve productivity.

Productivity also depends on many factors, such as the weather, regulatory quality, functioning of markets, input supply, efficiencies of processes, worker education, experience and skills, working hours, and the technologies embodied in different types of equipment.

Generally, construction productivity is lower than that of other sectors, such as manufacturing. Apart from the above factors, there is insufficient research and development (R & D) because of thin profit margins from competitive tenders. Further, the project team is temporarily assembled for the development, and this makes it difficult to improve productivity from *learning by doing* (Tan and Elias, 2000). Finally, unlike manufacturing, there are few *scale economies* in construction because every project is unique. A process $f(.)$ with output Q and inputs of capital (K) and labor (L) enjoys economies of scale if for any scalar ϕ,

$$Q^* = f(\phi K, \phi L) > \phi Q.$$

For example, if $\phi = 2$, doubling the inputs results in $Q^* > 2Q$. In other words, the output has more than doubled. If $Q^* = 2Q$, the process has constant returns to scale. Finally, if $Q^* < 2Q$, the process has diminishing returns to scale.

Productivity (A) may be measured as a residual using the index number approach, that is,

$$log\left(\frac{A_t}{A_{t-1}}\right) = log\left(\frac{Q_t}{Q_{t-1}}\right) - \alpha \, log\left(\frac{K_t}{K_{t-1}}\right) - \beta \, log\left(\frac{L_t}{L_{t-1}}\right)$$

where the parameters α and β are factor shares for capital and labor, respectively, t denotes time, and $log(.)$ is the natural logarithm function. The Tornqvist index (Tornqvist, 1936) uses dynamic factors shares; that is, α is the average of capital factor shares between adjacent time periods, and similarly for β. Factor shares are computed using rental rates for different types of capital and wages rates for different categories of workers.

Since for small changes

$$log\left(\frac{x + \Delta x}{x}\right) \approx \frac{\Delta x}{x},$$

the above equation may be approximated as

$$a_t = q_t - ak_t - \beta l_t$$

where a, q, k, and l represent rates of change. For example, if there is no change in capital input for a particular year ($k_t = 0$), $\alpha = 0.35$, $\beta = 0.65$, the labor force grows by 2% ($= l_t$) and construction output rises by 3% ($= q_t$), then productivity growth for that year is

$$a_t = 3\% - 0.35(0) - 0.65(2\%) = 1.7\%.$$

If constant returns to scale is assumed, then it is only necessary to compute β, which is easier than estimating α directly. Then, α is simply computed as $1 - \beta$. It is more difficult to obtain data on capital input because there are many different types of capital, such as computers, machines, equipment, vehicles, and buildings, with different vintages. There will also be difficulties in determining the different rates of depreciation and extent of capacity utilization.

In view of these difficulties, it is more common to measure only labor productivity, which is simply Q/L. This approach ignores capital input (K) and technical change (A). Even here, it is possible to mismeasure construction output Q and labor input L because of quality differences. If aggregate construction contract values (i.e. $\Sigma p_i Q_i$) are used as output, the change in output may be due to pure price changes. On the other hand, if physical units such as per unit of floor area constructed are used, there are many different types of buildings, making it difficult to compare productivity across firms and countries.

Submittals

Apart from the project plans, owners normally require the contractor to prepare a schedule of submittals that include shop drawings, product data, warranties, samples, test reports, and energy calculations. The contractor, subcontractors, fabricators, and manufacturers submit these items to the sponsor's project team for approval or review before using or installing them on site.

A typical process is for the subcontractor to submit shop drawings from a fabricator to the contractor. The contractor reviews the drawings, verifies the materials, measurements, and construction method, and stamps the drawing as "Reviewed" before passing them to the owner's project team. The team may approve, ask for corrections, or reject the drawings. It informs the contractor of the decision, and the latter conveys it to the subcontractor.

Designers prefer to "review" rather than "approve" shop drawings to reduce liability. They approve the design *intent* rather than the details, and argue that the contractor is responsible for making sure that all components fit together. A subcontractor or fabricator is expected to take final field measurements before fabrication.

The submittal process may be cumbersome if designers reject shop drawings or require revisions. A more efficient way is for the parties to meet, reduce the information required, or submit similar items as a single package.

Budget and cost systems

Based on the SoV in Table 11.1, the contractor will track *each* work package for any deviation in the budgeted value (now renamed as "cost") and take corrective actions. For the *entire project*, the contractor will also monitor the planned and actual costs. In Fig. 12.3, the vertical line is time now, and the difference between the planned and actual costs is the *cost variance*. The earned value (EV) is given by

$$EV = (\text{Planned cost}) \times (\% \text{ of scope completed}).$$

For example, if the planned cost of the project is $100 m and the scope of work done is 25% at the current time, then EV = $25 m. As shown in Fig. 12.3, the actual cost exceeds the EV, which is undesirable.

Managing quality

The contractor's quality plan is similar to that of the SPV (see Chapter 7); that is, it applies the principles of Total Quality Management (TQM), Quality Assurance (QA), and Quality Control (QC) at different levels (Table 12.4)

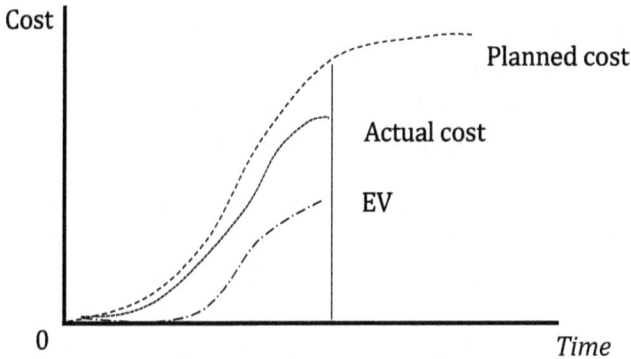

Fig. 12.3   Project cost control.

Table 12.4   Quality levels and approaches.

Level	Quality approach	Principles or tools
Company	TQM	Principles (see text)
Project	TQM	Principles (see text)
Process	QA	PDCA cycle, six sigma, re-engineering, failure mode and effect analysis, certification, incentives, audit, etc.
Product or service	QC	Cause and effect diagram, Pareto diagram, checklists, inspections, tests, samples, control charts, scatter diagram, etc.

The TQM principles include

- using quality as part of the corporate strategy to compete effectively;
- providing leadership for quality standards and improvement;
- providing an effective organization structure for quality improvement;
- adopting a systems or holistic approach to quality;
- adopting a customer focus and thorough understanding of customer requirements;
- encouraging employee involvement and empowerment;
- implementing a process approach to quality;

- striving for continuous improvement or *kaizen* (Imai, 2012);
- implementing productive maintenance;
- applying evidence-based decision-making; and
- cultivating win-win supplier and subcontractor relations.

QA focuses on the process level, such as improving a contractor's bid preparation, tender process, project planning, and so on. Finally, at the product or service level, quality control aims to reduce defects and variability.

There are many tools for improving quality. For example, in the Plan-Do-Check-Act (PDCA) cycle, the planning phase consists of

- defining the process;
- developing the flow chart; and
- documenting the process.

The "do" phase consists of

- evaluating the process;
- ascertaining the causes, such as design, process, equipment, materials, finishing, inspection, testing, packaging, or worker; and
- finding new solutions.

The proposed solutions are then "checked" for effectiveness before "acting" on it, that is, implementation and evaluation. Readers can read up books on TQM for other tools (e.g. (Kiran, 2016)).

Some countries have systems for evaluating the contractor's output quality, such as Singapore's Construction Quality Assessment System (CONQUAS), which was developed in 1989. It assesses the structural, architectural, mechanical, and electrical works in building projects. Contractors with good CONQUAS scores have a competitive advantage when tendering for projects.

Managing safety and health

The contractor's safety and health (SH) plan consists of

- a general policy statement of management commitment and that everyone is responsible for SH;

- an organization structure for assigning responsibility, monitoring, and reporting;
- a system for identifying hazards;
- SH standards, e.g. fall protection, demolition, ergonomics, preventing heat stress, and protective gear;
- mitigation measures;
- briefing, communication, signage, and training;
- site SH rules;
- incident management; and
- SH requirements for subcontractors and other external parties.

In other words, SH thinking requires a holistic approach to planning and execution (Moran, 2003; Goh, 2021).

The contractor's SH plan is not a stand-alone document. It should integrate the SH plans of the owner, subcontractors, suppliers, and other stakeholders. For example, the supplier of certain equipment may require that certain precautions be taken during installation.

The common health issues in a construction site include dust, mental ill-health, stress, heat, noise, poor sanitation, chemical exposure, working in confined spaces, over-exertion, and inappropriate accommodation.

The record of safety in construction is not good. For example, according to the US Bureau of Statistics, nearly 1 in 5 workplace deaths occurred in the construction industry in 2021, and mostly through falls, slips, and trips (*The Economics Daily*, 2023). Since many countries have similar construction fatality rates, more can be done to improve construction safety and health through training, proper risk assessment, and the use of appropriate incentives.

Managing scope

Scope changes are inevitable in large projects, and they must be within the original scope of the contract. For example, a change from a school to a prison is beyond the scope of a "school." Scope creep, or small changes in scope, can occur through marginal design changes.

Owners may require scope changes because of business considerations, such as altering production capacity because of changing product demand or there is a change in the building code or regulation. The contractor may also initiate scope changes if there are differing site conditions, defective specifications, or a specified product is not available. Scope changes are common in software development projects because there are many users or departments, and each wants to tailor the software to their current business procedures.

A *differing site condition* is one that differs from the contract documents or one which the contractor does not reasonably expect. Since it does not make sense for each bidder to conduct soil tests on the site prior to tender, the sponsor normally supplies the soil test data "for information only," implying that bidders should conduct their due diligence. However, given the lack of site access and the short time for tender, contractors price their bids based on the owner-supplied data. Hence, when site conditions differ from the soil test data, the contractor expects the owner to issue a *variation order* (VO) or *change order* (CO) to instruct the contractor to make changes. Generally, the owner takes this risk because, as explained earlier, it is not reasonable to expect the contractor to conduct detailed soil tests before the tender. If the owner shifts the risk to the contractor, the latter will include a high contingency sum and stands to gain if the differing conditions do not occur. The tender may also attract a risky bidder who submits a low contingency sum.

However, if the site conditions are expected from the owner's soil test data, and the contractor did not exercise due diligence at the time of bid, the owner does not need to issue a VO. It is assumed that owners do not intentionally hide site information to avoid paying for extra work.

The VO process starts with the owner's project team or the contractor. In the former case, the owner's side requires scope changes. In the latter case, the contractor discovers the need for a VO. Both sides must agree on the cost and schedule impact of the proposed VO. If there is disagreement, one option is for the initiator to withdraw the VO. The contractor may revise or pursue as submitted, escalating it to a *claim*; that is, an unsettled benefit that a contractor believes he is entitled. Both sides should use a VO log to track changes (Table 12.5).

A VO is a formal written request. It is possible for a sponsor to informally request for scope changes. It is called a *constructive change* and, in the absence of a formal request, can lead to disputes.

Table 12.5   VO log.

VO proposal	Description	Submit	Outcome				Amount ($)
			Approved	Rejected	Revised	VO No.	

Managing stakeholders

Just like the grantor and sponsor, the contractor needs a plan to manage stakeholders (Table 12.6). Stakeholder management is about understanding their concerns (interests) and taking steps to resolve these issues. Stakeholders have differing power depending on their resources and

Table 12.6   Stakeholder management plan.

Stakeholder	Interest	Priority	Modes of engagement
Grantor	Project completion	Keep informed	Through SPV
SPV	Project completion	Manage closely	Project meetings
SPV's lender	Project completion	Keep informed	Through SPV
Operator	Project completion	Keep informed	Through SPV
Politicians	Self or public interest	Keep informed	Periodic updates
Regulators	Self or public interest	Manage closely	Meetings
Subcontractors	Prompt payment, good working relations	Manage closely	Project meetings
Suppliers	Prompt payment, good working relations	Manage closely	Project meetings
Customers/users	Project completion	Manage closely	Meetings
Community	Traffic, noise, etc.	Keep informed	Periodic updates
Mass media	General public issues	Keep informed	Periodic updates
Pressure groups	Social, environmental, and other issues	Keep informed	Periodic updates
Internal	Project completion	Manage closely	Internal modes

organization. Obviously, key stakeholders need to be managed closely. Note that the stakeholder management plan in Table 12.6 is from the perspective of the contractor.

Pro-community discourses or narratives play an important role for the contractor. Stakeholder engagement needs to be inclusive and participatory; to do so, good feedback needs to be considered and acted upon.

Commissioning plan

The contractor integrates the commissioning with the sponsor's plan (see Chapter 10) on the following:

- commissioning team;
- systems to be commissioned;
- risk assessment;
- special resources required;
- commissioning schedule;
- acceptance criteria;
- documentation;
- training of operatives; and
- issuance of certificates.

Recall that the systems to be commissioned include power, lighting, HVAC, refrigeration, plumbing, fire, communications, computer, security, vertical transport, waste disposal, building envelope, structural, and special systems.

Managing risks

Recall from Chapter 3 that the grantor conducts risk assessment during economic appraisal using both quantitative and qualitative techniques, such as sensitivity analysis, Monte Carlo simulation, risk-adjust discount rates, and project options. The grantor normally assumes political and regulatory risks and allocates commercial risks to the SPV. In turn, the SPV allocates construction or completion risks to the contractor. In addition, both parties may share *force majeure* risks that are beyond their control. Hence, the contractor's risk management plan focuses on project planning and execution.

The risk management process consists of

- identifying the risks;
- prioritizing the risks by assessing likelihoods and impacts;
- developing mitigation measures; and
- monitoring and review.

For the contractor, the main risks relate to (Greiman, 2013)

- unclear or misunderstanding of the project scope;
- unexpected site conditions;
- cost, e.g. poor bid estimate, unforeseen cost changes, and failure to keep within budget;
- time, e.g. poor estimates of task durations, sequencing, and coordination;
- quality, e.g. defects;
- safety;
- labor issues;
- unfamiliarity with new technologies;
- unfamiliarity with local conditions or practices;
- design errors and omissions, if the contractor is also responsible for design;
- procurement, e.g. non-performance of suppliers and subcontractors;
- cash flow, such as payment issues; and
- difficulties in settling disputes, including insurance claims.

For each risk, it is possible to assess the consequence or risk exposure, which is defined as the likelihood of occurrence × impact. Usually, a 5-point scale is used to rate the likelihood and impact of a risk event. Risk events are then ranked by their consequences, as illustrated in Table 3.8.

Recall from Table 3.9 that there are a number of risk mitigation strategies. Finally, the contractor monitors the risks and conducts periodic risk reviews. As the project progresses, risks may change.

Managing subcontractors and suppliers

The contactor's procurement plan deals with the appointment of subcontractors and purchases of goods and services. Contractors use their network of subcontractors and suppliers. For purchases, the site or engineering team makes a proposal to the procurement department. The latter

- approves the proposal;
- prepares the purchase contract;
- searches for potential suppliers;
- conducts the tender;
- analyzes the bids;
- prepares the purchase order; and
- monitors progress, such as the requirement for test certificates and shipping.

For imports, the standard shipping terms are free on board (FOB) and cost, insurance and freight (CIF). These are international commerce terms (incoterms). FOB is cheaper as the seller is responsible only to the point where the item is loaded onto a ship. Once the ship sails, the buyer assumes the costs and liabilities. In a CIF arrangement, the seller is responsible until the buyer receives the item, that is, up to the project site or factory gate. The preferred arrangement is likely to depend on whether the seller is familiar with local customs, taxes, and transport conditions.

The contractor needs to manage the subcontractors working in the project. The contractor should be included as additional insured in the subcontractor's insurance policy to protect against third-party claims. Effective coordination of work with subcontractors requires close supervision, regular meetings, timely release of information ahead of schedule, addressing safety and health concerns, and proper documentation.

Finally, subcontractors need to be paid promptly when they submit their payment requests to the contractor. The latter consolidates the requests and submits them to the owner for payment. The owner's project team, and possibly the lender's consultant, will then measure and certify actual progress. In the past, contractors pay subcontractors only when they are paid by owners or have the funds and are willing to do so.

The "pay when paid" approach is no longer applicable in many countries. Some countries have introduced the Security of Payment Act to expedite payment to subcontractors (Lim, 2020). Depending on the jurisdiction, the contractor has 2–3 weeks to respond to a claim for payment. If the subcontractor is not satisfied with a partial payment or there is no response from the contractor, she may apply for adjudication. This approach is faster and less costly than litigation. Often, the contractor offers a partial payment because work has not been completed or there are defects.

Inspection plan

The owner's project team will want to inspect certain aspects of the construction. The inspection plan sorts out the inspection protocol. It comprises the following:

- the items to inspect;
- inspector's qualifications;
- limits to inspector authority;
- testing procedures;
- hold points and documentation;
- acceptance procedures;
- procedure for resolving disputes;
- inspection reports; and
- issuance of the inspection certificate.

A "hold point" is a point where the contractor cannot cover the work without proper inspection, such as in pouring concrete to erect a building column. The consulting engineer or her representative will want to see that the reinforcement bars are in place before concrete is poured.

Documentation plan

The contractor's documentation plan shows the documents and folders at each stage of the project cycle (Table 12.7).

Many of these documents are stored electronically in databases.

Table 12.7   Contractor's documentation plan.

Project stage	Folders/Documents
Tender	Tender documents, subcontracts
Mobilization	Permits and approvals, baseline plans, bonds, insurance
Construction	Cost and status reports, submittals, variation orders, claims and disputes, tests and inspections, progress payments, work certification, progress reports, procurement, labor records, visitor logs, minutes of meetings, correspondences, safety records, maintenance logs
Project close-out	Commissioning test results, punch list, certificates, final documents, training of operatives, project evaluation, post-occupancy audit

Communications plan

The contractor's communications plan shows the mode of communication, purpose, attendees, chair or author(s), means of dissemination, and frequency (Table 12.8).

The communications plan is not only about information exchange. It is an opportunity to build trust, generate support, manage change and resistance, mitigate conflict, and facilitate stakeholders' satisfaction (Plowman and Diffendal, 2020).

Progress reports

The contractor's project team usually submits two reports. The project manager submits a weekly or monthly progress report to senior management of the construction company on the following:

- scope changes and VOs;
- cost changes and analysis;
- schedule changes and analysis;
- subcontractors;
- suppliers; and
- issues.

In turn, the contractor submits a project progress report to the sponsor based on inputs from the project manager.

Table 12.8   Contractor's communications plan.

Mode	Purpose	Attendees	Chair/author	Means	Frequency
Kick-off meeting	Introduce project	Project team, subcontractors	PM	Meeting	Start of project
Status report	Update management	—	PM	Email	Monthly
Project team meetings	Review progress and take action	Project team	PM	Meeting	Weekly
Project Advisory Group meeting	Update and resolve issues before escalating to sponsor	PM, Project Advisory Group	PM	Meeting	Monthly
Sponsor meeting	Update and resolve issues	Sponsor's project team, contractor's project team	PM	Meeting	Monthly
External stakeholders' meeting	Inform, resolve issues	Project team, external stakeholders	PM	Meeting	Ad hoc
Debrief	Post-project review	Project team, subcontractors	PM	Meeting	End of project
Debrief	Post-project review	Project team, sponsor	Sponsor's PM	Meeting	End of project

The superintendent submits a daily report (also called daily logs or diaries) to the project manager on

- weather;
- general field progress;
- presence of subcontractors and their workforce by craft and equipment;
- deliveries;
- occurrences and reasons;
- visitors; and
- issues.

These reports may include progress images. They may also be used for settling future disputes.

Claims and disputes

Recall that a claim is an unsettled benefit a contractor believes she is entitled. The causes of claims include (Netscher, 2016)

- differing site conditions;
- relocation of work by the sponsor;
- lack of site access;
- design changes or insufficient detail;
- defective specifications;
- design errors and omissions;
- acceleration of work as directed by the sponsor;
- unnecessary restrictions on methods;
- interrupted work;
- sponsor furnishes equipment or materials late or in poor condition;
- late inspections, reviews, or approvals;
- *force majeure* events;
- late payment;
- failure to agree on VO pricing;
- variations in quantities in unit price contracts;
- rejection of requested substitutions; and
- improper work rejection.

A claim may escalate into a *dispute*. While claims are in progress, the contractor cannot stop work. Obviously, the sponsor's project team may challenge a claim, which is why claims require evidence and proper documentation (Rubin *et al.*, 1999).

There are many ways to reduce or avoid claims, such as equitable risk allocation, better information, appropriate contract clauses, effective project management, partnering, and proper documentation (Godfrey, 2006; Levin, 2016).

Claims may be resolved through *negotiation*, failing which both parties may seek third-party *reconciliation* where the decision of the reconciliator is not binding. If this does not solve the problem, the next stage is for both parties to appoint a *mediator*. Some projects use a dispute review board comprising three members to make the recommendations. Both parties appoint the board members prior to the start of construction.

If the desire is for binding decisions rather than just recommendations, there are several options. *Adjudication* is a statutory process of hearing disputes, such as payment issues between the contractor and subcontractor, under the Security of Payment Act. In *arbitration*, both parties appoint an arbitrator who then makes a binding decision. The most costly option is *litigation*, where parties settle it in court. It is the last resort. In a *default judgment*, the defendant fails to enter appearance or file her defense. A second type of decision is a *summary judgment*, where the plaintiff can show that the defendant has no real defense. The defendant will try to show that this is not the case and claim for a full trial.

Apart from differing site conditions, another common area of dispute is delays along the critical path, and the project cannot finish on time (Nagata *et al.*, 2017). If the contractor causes the non-excusable delay, the sponsor can impose liquidated damages. Similarly, if the sponsor causes the non-excusable delay, the contractor can expect an extension of time and possible compensation. If both parties cause an overlapping or concurrent delay, the resolution depends on contract clauses. If the contract is silent on this matter, then there is usually no compensation, but the contractor may be given an extension of time. For example, if the sponsor is late in providing the drawings and this overlaps with the contractor's delay by his subcontractor, then the contractor may ask for an extension of time.

Finally, there is the serious possibility of early termination of the construction contract because of the contractor's poor performance, the sponsor runs out of funds, there is a *force majeure* event, or there is frustration of contract. A frustration of contract occurs if either party is unable to perform because of unforeseen circumstances. For example, if the contractor does not have access to the site, he cannot perform, and there is no breach of contract. The sponsor must seek the lender's approval to terminate the construction contract early, and the three parties will not take this lightly. The lender may exercise her step-in rights under the Direct Agreement, as discussed in Chapter 7.

References

Godfrey, K. (2006) *Partnering in design and construction.* New York: McGraw-Hill.

Goh, Y. M. (2021) *Introduction to workplace safety and health management.* Singapore: World Scientific Publishing.

Goldratt, E. (1999) *Theory of constraints.* Great Barrington: North River Press.

Greiman, V. (2013) *Megaproject management: Lessons on risk and project management from the Big Dig.* New York: Wiley.

Imai, M. (2012) *Gemba kaizen.* New York: McGraw-Hill.

Kiran, D. (2016) *Total quality management.* London: Butterworth-Heinemann.

Levin, P. (Ed.) (2016) *Construction contract claims, changes, and dispute resolution.* Virginia: ASCE.

Lim, P. (2020) *Contract administration and procurement in the Singapore construction industry.* Singapore: World Scientific Publishing.

Moran, M. (2003) *Construction safety handbook.* Maryland: ABS Consulting.

Nagata, M., Manginelli, W., Lowe, S., and Trauner, T. (2017) *Construction delays: Understanding them clearly, analyzing them correctly.* London: Elsevier.

Netscher, P. (2016) *Construction claims: A short guide for contractors.* Perth: Panet Publications.

Plowman, C. and Diffendal, J. (2020) *Project communications.* New York: Business Expert Press.

Rubin, R., Fairweather, V., and Guy, S. (1999) *Construction claims.* New York: Wiley.

Tan, W. (2010) The elasticity of capital-labor substitution in Singapore construction. *Construction Management and Economics*, **14**(6), 537–42.

Tan, W. and Elias, Y. (2000) Learning by doing in Singapore construction. *Journal of Construction Research*, **2**, 151–58.

The Economics Daily (2023) Construction deaths, due to falls, slips, and trips increased 5.9 percent in 2021. 1 May.

Tornqvist, L. (1936) The Bank of Finland's consumption price index. *Bank of Finland Monthly Bulletin*, **10**, 1–8.

CHAPTER 13

Project close-out

Close-out activities

As the construction nears completion, the various parties need to manage and coordinate the following close-out activities (Mincks and Johnston, 2017):

- conduct start-up and testing;
- rectify punch lists;
- issue a Certificate of Substantial Completion;
- settle final payment;
- hand over documents and materials;
- apply for a Certificate of Occupancy;
- prepare for operation and maintenance (O & M);
- release resources; and
- evaluate project performance.

These activities are discussed below.

Start-up and testing

The contractor will arrange for the commissioning, startup, and testing of the various systems in accordance with the Commissioning Plan (see Chapter 12).

Depending on the system to the tested, the attendees may include the sponsor, the project team, contractor, commissioning experts, relevant authorities, subcontractors, suppliers, and manufacturers.

Punch lists

The contractor will collate a preliminary punch list of items for subcontractors and suppliers to rectify. When these works have been completed, the contractor will inform the owner's project team of the work completion. The project team will then conduct a pre-final inspection with the contractor and subcontractors and issue a punch list for the contractor to rectify.

Certificate of Substantial Completion

After the rectification of defects, the project team will conduct a final inspection. If the works have been rectified, it will issue a Certificate of Substantial Completion to the contractor. In some countries, it is called the certificate of practical completion. The issuance starts the clock for the defects liability period (DLP), warranties, and liquidated damages for project delay, if any. The DLP is usually a year. The owner will take over the responsibilities for property and liability insurance, routine maintenance, utilities, and security.

When the minor works are also completed, the designers will issue a Certificate of Final Completion.

Final payment

After the issuance of the Certificate of Substantial Completion, the contractor and owner will settle all outstanding payments and claims, and the latter will release half the retainage (e.g. 2.5% of 5%). The other half will be released after the DLP when all defects have been satisfactorily rectified.

The settlement of final payment can be contentious if the contractor has a large number of claims. Often, both parties take a give-and-take approach, especially if the project does not have major issues. Sometimes, they have to carefully document the issues and prepare for major disputes.

Handing over of documents and materials

The contractor will hand over the following contractual items to the owner:

- building user guide for the various building systems;
- testing and commission results;

- certificates and warranties;
- as-built drawings;
- keys;
- spare materials and parts;
- copies of statutory permits, licenses, and approvals;
- Building Information Modeling (BIM) files from the common data environment;
- confirmation of payment and lien releases; and
- consent of surety.

The contract may stipulate additional items to hand over, such as software to operate certain systems. If there are energy and other performance requirements or targets, such as Leadership in Energy and Environmental Design (LEED), Green Building, Building Wellness, or carbon credit certification, the relevant authorities, designers, or consultants will issue the certificates (Kibert, 2016).

The contractor provides a confirmation letter to the owner that all payments have been made, and subcontractors will need to provide lien releases or waivers, which are undertakings not to put a lien on the property. In some countries, a subcontractor who has not been paid may put a lien on the property so that it cannot be sold until the debt has been settled.

The consent of the surety to terminate the performance bond is required so that the contractor can stop paying the premium. If the contractor does not rectify the defects during the DLP, the owner will deduct the monies from the retainage.

Certificate of Occupancy

The owner needs to apply for a Certificate of Occupancy from the relevant building authority. It certifies the completion of building works in accordance with the earlier development plans.

In many cases, not all works are complete, and the owner applies for a Temporary Certificate of Occupancy to allow tenants and users to occupy the building. Once the works are completed, the owner will apply for a Final Certificate of Occupancy.

These certificates go by different names in different countries, such as Temporary Occupation Permit, Temporary Occupancy Permit, and Certificate of Statutory Completion.

Preparation for operation and maintenance

To prepare for the operation and maintenance (O & M) phase, the contractor may be contractually obliged to provide training for operatives of the various systems.

Release of resources

The contractor will begin to redeploy his resources to other projects as the construction nears completion. The project office will be progressively downsized until a small team remains during the DLP.

Evaluation of performance

Project evaluation may consist of two sessions, one for direct project participants after the final payment and the other for users a few months after occupation (Preiser *et al.*, 2016). The purpose of these sessions is to gather the following:

- feedback from the owner's project team;
- where the contractor did well;
- areas for improvement; and
- feedback from subcontractors and suppliers.

The owner may also conduct internal evaluations of performance as well as external feedback on the social, economic, and environmental outcomes of the project.

The parties will need to access, retain, protect, develop, and share the critical and often tacit or experiential knowledge gained from the project (O'Dell and Hubert, 2011). The sharing may take the form of mentoring, networking among communities of practice, and transfer of best practices for future projects.

Feedback may not be genuine if the parties fear repercussions for voicing criticisms. There needs to be a balance of positives and negatives, tact, and a real interest to improve. Otherwise, the feedback may not be specific or are of minor consequences (Stone and Heen, 2015).

References

Kibert, C. (2016) *Sustainable construction: Green building design and delivery.* New York: Wiley.

Mincks, W. and Johnston, H. (2017) *Construction jobsite management.* Boston: Cengage Learning.

O'Dell, C. and Hubert, C. (2011) *The new edge in knowledge: How knowledge management is changing the way we do business.* New York: Wiley.

Preiser, W., White, E., and Rabinowitz, H. (2016) *Post-occupancy evaluation.* London: Routledge.

Stone, D. and Heen, S. (2015). *Thanks for the feedback: The science and art of receiving feedback well.* New York: Penguin.

CHAPTER 14

Operation and maintenance

Asset management

The operation and maintenance (O & M) phase is the longest period in a public–private partnership (PPP) contract, often lasting throughout the asset's life and is usually around 25–35 years.

There are two types of asset management — financial and physical. Financial asset management deals with managing the funds of individual and institutional investors by investing in financial assets such as bonds and equities (Fabozzi and Fabozzi, 2020). The goal is to maximize the value of the portfolio and, hence, shareholder value for any given level of risk.

Physical asset management, which concerns us here, deals with managing the fixed assets of the special purpose vehicle (SPV), such as buildings, vehicles, furniture, machinery, equipment, and computers (Hastings, 2022). The goal is to improve the efficiency and performance of these assets to drive organizational objectives. For the private sector, the overriding concern is profitability. The public organization's concern is value for money for infrastructure and other public assets.

Real Estate Investment Trusts (REITs) raise funds through equity and debt to invest in real estate rather than financial assets to maximize the rental return (Scarrett and Wilcox, 2018). Hence, it is a form of physical asset management. REITs are landlords of commercial buildings, offices, and industrial properties, and these manage the assets with a strong financial motive.

Both types of asset management also deal with regulatory and contractual compliance. In summary, the goal of asset management is to buy, operate, maintain, and sell assets efficiently at an acceptable level of risk while satisfying contractual and regulatory requirements.

Strategic asset management

With the above goals in mind, the SPV manages its assets by first developing a strategic asset management plan (SAMP) (Table 14.1) or asset management action plan (United Nations Department of Economic and Social Affairs, 2021).

Table 14.1   Strategic asset management plan.

Asset	Purchase price	Date of purchase	Current value	LCC	Required service standards	Financing
1						
2						
...						
N						

An SAMP is developed column by column, starting with a list of the key assets it owns. For each asset, the information required are the purchase price, date of purchase, current value, life-cycle cost (LCC; see Chapter 1), the required service standards, and ways to finance it. The financing includes the initial investment and the operating, maintenance, replacement, and disposal costs. More columns may be added as desired, such as repair and replacement dates and maintenance strategies.

The required service standards vary for each type of asset. For instance, in the case of elevators, the service standards include capacity, passenger load, and speed. Service standards for cleanliness of general premises are based on minimum scores on a rating scale. Similarly, service standards for buses are based on mileage, frequency of service, and rate of accidents.

From the SAMP, the SPV may decide to operate the assets and manage the building services. More commonly, it will focus on managing the PPP contract and outsource these activities to an operator and facilities manager, respectively. In some cases, both the operator and facilities manager are also project sponsors, and there may be conflict of interests.

Operation and maintenance contract

The contract for the operation and maintenance (O & M) of an infrastructure asset depends on the type of asset. For example, for a train system, the operation includes scheduling, ticketing, signaling, maintenance, and so on. The O & M contract for a train operator deals with

- O & M goals;
- appointment and term of contract;
- representations and warranties;
- scope of works;
- regulatory compliance;
- responsibilities and rights of owner;
- representative of each party;
- information, reporting, audit, and records;
- O & M program and budgets;
- service standards, fees, incentives, and penalties;
- responsibilities for costs and expenses, including capital expenditure, limits on spending, and specialized maintenance;
- *force majeure* arrangements;
- defaults, remedies, and termination;
- indemnity;
- limitation of liability;
- insurance;
- assignment of benefits;
- interface management, e.g. shared railway interchanges;
- use of computer software for O & M;
- confidentiality;
- emergencies; and
- dispute resolution.

Since these activities involve considerable investment for the operator, the term for an O & M rail contract is about 10–15 years.

Facilities management contract

The facilities management (FM) contract is about managing the facilities on behalf of the SPV. The main goals in FM are to optimize asset performance, reduce operating costs, extend asset life cycles, and improve user wellness. In the context of a rail system, the FM contract usually covers the provision of building services to the railway stations. The facilities manager does not operate the trains.

FM services may be categorized as hard and soft. The hard FM services cover utilities, heating, ventilation, and air conditioning (HVAC), elevators, drainage, fire systems, building façades, interior architecture as well as computer and telecommunication networks.

The soft FM services include solid waste management, cleaning, minor repairs, security, car park management, hosting and catering, pest control, landscaping, and office relocations.

The contract clauses for FM are similar to that of the O & M contract. The SPV's main FM concerns are cost-efficiency and reliability. Unlike O & M contracts, the term for FM contracts is shorter (e.g. three years) because the facilities manager does not incur much upfront investment.

Principles of maintenance

The operator or facilities manager should develop a reliability-centered maintenance (RCM) program that comprises a mixture of corrective, preventive, predictive, or condition-based maintenance. The performance of such a program may be measured in terms of operational efficiency, reliability, and safety. The key components are identified for maintenance to achieve high reliability (Bloom, 2005).

Corrective maintenance is usually applied to small and non-critical items that have failed. For instance, light bulbs are often replaced only after failure. The impact on operations is negligible. Corrective maintenance is reactive and is also called breakdown or run-to-failure maintenance.

Preventive maintenance is carried out at regular intervals, irrespective of the condition of the asset. For example, air conditioners are often maintained at quarterly intervals to balance the cost and service quality. Preventive maintenance may not be cost-effective because it is applied without regard to asset condition. For example, if the air conditioners in a school are maintained on a quarterly basis, the asset conditions may vary considerably depending on usage.

Predictive maintenance is used for maintaining key assets where failure has serious consequences. The approach uses operational data, artificial intelligence, and diagnostic equipment to better predict asset stress. The asset is then maintained or repaired at a sufficient buffer just before failure.

In condition-based maintenance, the condition of the asset determines when it should be maintained. Hence, predictive and condition maintenance are similar; both approaches are based on the condition of the asset.

The difference between preventive, predictive, and condition maintenance narrows if the maintenance interval correlates highly with the condition of the asset. Hence, the condition of an asset is a key determinant of maintenance strategy.

The condition of an asset is determined from data and statistical analyses. The common statistics are the mean and standard deviation of the life of an asset. With this knowledge, it is possible to predict the remaining useful life (RUL) of an asset. In Fig. 14.1, there are three identical assets (e.g. pumps), with t being the current time, T is failure time, C is the current level of performance, and F denoting performance failure. There are three failure times (T_1, T_2, and T_3) shown as dotted lines but are not labeled to avoid clutter. If there are a large number of identical assets, it is possible find the mean failure time and its standard deviation.

There are different predictors of the conditions of an asset. For example, the conditions of a road include distress, roughness, deformation,

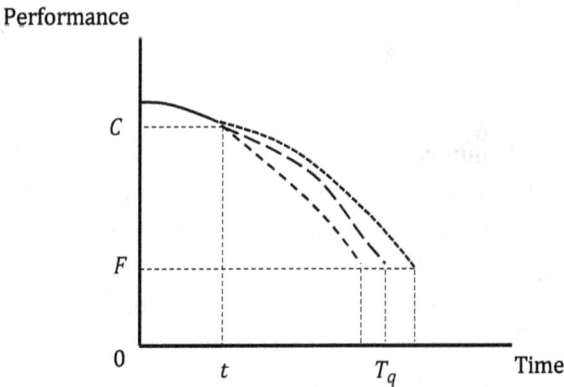

Fig. 14.1   Estimating the remaining useful life of an asset.

faults, and friction. For any physical asset, the conditions may be detected using techniques such as

- visual analysis;
- vibration analysis;
- oil analysis;
- infrared thermography;
- ultrasound technology;
- artificial intelligence (AI); and
- digital twins.

For assets with obvious physical deterioration, such as large potholes on a road, visual inspection may be used. Drones have largely replaced the need for an inspector to physically do the checking.

Vibration analysis is used to detect problems with rotating equipment, such as loose parts, poor alignment, bearing faults, or unbalanced components (Goldman, 1999). The general process is to sample a few blocks of data in the time domain (e.g. 4 blocks of 10 seconds each) and then use Fast Fourier Transform (FFT) or any suitable method to transform each block to the frequency domain comprising sine and cosine functions. The spectra of these blocks are then averaged. Peaks in the mean spectrum are then compared to the theoretical frequencies to identify possible faults (Myklestad, 2018). The transformation from input time domain $x(t)$ to output frequency domain $X(\omega)$ is given by the Fourier transform

$$X(\omega) = \sum_{t=-\infty}^{\infty} x(t) e^{-j\omega t}$$

where e is Euler's number and $j^2 = -1$; that is, j is an imaginary number called the complex operator. The use of the j operator obviates the need to write and estimate two separate sine and cosine equations.

In oil analysis, samples of the lubricant are sent for a large number of laboratory tests to determine the oil condition, identify contaminants, and analyze wear debris from machinery (Hunt and Evans, 2008). The results may indicate the following: a need to change the oil, the machinery is operating inefficiently, and/or it may soon fail.

Infrared thermography uses a thermal imager or infrared camera to detect the infrared radiation emitted from an object (Lanzoni, 2015). The heat distribution in different colors (thermograph) is used to diagnose the condition of the asset. For example, it can detect heat in machinery caused by excessive friction. Thermographs are also used to detect voids (laminations) and water infiltration in concrete structures, among many other applications.

Ultrasound technology uses piezoelectric crystals in the transducer to generate ultrasound waves to detect internal conditions, such as bearing faults. It is a method for non-destructive testing (NDT) of materials and machines (Blitz and Simpson, 1995). The reflected echoes are converted to electrical signals that are then transformed into images and analyzed. The ultrasound probe has many applicants, such as spotting gas leaks, identifying faulty bearings, checking valves, testing the structural properties of concrete, and ascertaining lubricant levels.

Artificial intelligence

Unlike the above technologies, AI is based on data. AI is an umbrella term for expert systems, machine learning, and deep learning. It is "artificial" because it is based on the computer, not humans, and it is "intelligence" because machines (computers) are taught to think instead of following the rules.

Early work on AI started in the 1950s, with expert systems or the use of experts to devise rules or "expert advice" for computers to follow (Durkin, 1993). The approach does not work well for less structured problems. For example, there are too many rules and exceptions in translating a language.

The new approach, called machine learning (ML), lets the computers learn the rules from data using statistical tools. Deep learning is a subset of ML and is based on neural networks (Kelleher, 2019).

The application of ML to predictive maintenance uses sensors on a machine to collect data (Prado, 2020). From the data, the machine will be able to detect data patterns or correlations using statistical tools (Tan, 2022). In Fig. 14.2, data are collected from two sensors on variables X and Y.

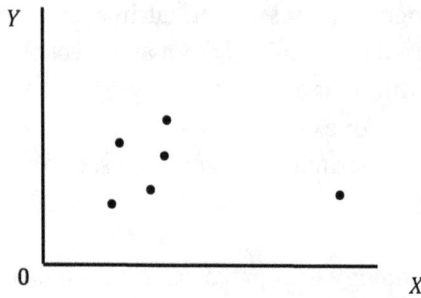

Fig. 14.2   Illustration of outlying data points.

There is an outlier with an abnormally large X value, indicating possible malfunction.

Digital twin

A digital twin is an up-to-date representation of a functioning asset, such as a pump, engine, or car. Hence, the digital twin is as old as the asset and should have the same condition. The representation may be a data-driven model or a physics-based model (Asch, 2022).

An example of a data-driven model is in Fig. 14.1, where the current state (performance) of the asset is known ("tuned") from data collected from sensors in real time. The traditional approach is to predict the performance of the asset based on current data. The digital twin approach differs by simulating how the asset will perform when different types of faults occur.

A physics-based digital twin model is based on the actual physical workings of the asset, such as how a jet engine works. By collecting data from an existing engine, it is possible to use the digital twin to analyze and predict engine problems. For example, Tesla has a digital twin of every car it sells. It then collects data from the car under different operating conditions in different countries. The data are then used to remotely optimize its performance.

Asset enhancement

Asset enhancement initiatives (AEI) refer to asset upgrades that increase net lettable area (NLA), lower costs, or improve asset value. An increase in NLA raises revenue. It is also possible to lower operating costs, such as installing energy-efficient devices or solar panels. Finally, many upgrades improve asset value, such as refurbishing restrooms, escalators, dining areas, and lifts.

AEI strategies are often linked with ESG strategies (see Chapter 1) to enhance value by improving the environment as well as the organization's social responsibilities, and governance.

Realizing the asset value

Sponsors need not wait until the end of the PPP contract to realize asset value. A sponsor may sell her shares to existing shareholders or an external party, such as an infrastructure fund, depending on the terms of the shareholders' agreement. Often, this requires the approval of the lender and grantor to ensure that the SPV will continue to function effectively.

Alternatively, the sponsors may decide to sell the SPV collectively to a third party, which may be an infrastructure group, or fund with expertise in running the project. Sponsors have different reasons for selling out, such as profitability, change in priority, perceptions of new risks, cash flow considerations, or difficulties working together. The infrastructure group or fund may securitize the asset by issuing new shares to other investors. In this way, it can raise funds by pooling infrastructure assets and issuing stapled securities. The pool is often well-diversified to reduce risks and achieve stable returns.

Infrastructure groups or funds may also provide project loans or purchase the project bonds. In addition, they may also invest indirectly, such as by investing in other infrastructure funds.

References

Asch, M. (2022) *A toolbox for digital twins*. Philadelphia: SIAM.

Blitz, J. and Simpson, G. (1995) *Ultrasonic methods of non-destructive testing*. London: Chapman and Hall.

Bloom, N. (2005) *Reliability-centered maintenance*. New York: McGraw-Hill.

Durkin, J. (1993) *Expert systems: Design and development*. New York: Macmillan.

Fabozzi, F. and Fabozzi, F. (2020) *Fundamentals of institutional asset management*. Singapore: World Scientific Publishing.

Goldman, S. (1999) *Vibration spectrum analysis.* South Norwalk: Industrial Press.

Hastings, J. (2022) *Physical asset management.* Berlin: Springer.

Hunt, T. and Evans, J. (2008) *Oil analysis handbook.* Chipping Norton: Coxmoor.

Kelleher, J. (2019) *Deep learning.* Cambridge: MIT Press.

Lanzoni, D. (2015) *Infrared thermography.* South Carolina: CreatSpace.

Myklestad, N. (2018) *Fundamentals of vibration analysis.* London: Blackwell.

Prado, M. (2020) *Machine learning for asset managers.* London: Cambridge University Press.

Scarrett, D. and Wilcox, J. (2018) *Property asset management.* London: Routledge.

Tan, W. (2022) *Research methods: A practical guide for students and researchers.* Singapore: World Scientific Publishing.

United Nations Department of Economic and Social Affairs (2021) *Managing infrastructure assets for sustainable development.* Geneva: UN DESA.

CHAPTER 15

Handing over

Early termination

A public–private partnership (PPP) project may be terminated by mutual consent or by either party. The reasons for termination include

- insolvency;
- fraud;
- breach of contract;
- a *force majeure* event;
- policy change;
- perceived operational inefficiencies;
- better opportunities;
- changed circumstances;
- bad faith;
- change of government; or
- perceived high profitability.

Terminating a project in bad faith means that the grantor has no intention of building the infrastructure; that is, a political stunt to announce a tender for the project because of an upcoming election. Sometimes, a new government may wish to terminate a PPP contract that it thinks may not have been awarded in a transparent manner by the previous administration, or it has been stuck with a high-cost operator. It is also possible for the grantor to terminate the project because of the perceived high profitability that irates paying users and other stakeholders. There may be the perception that the grantor made a gross error in estimating the profitability of the project.

If the special purpose vehicle (SPV) terminates the project during construction, the grantor will have to find another party to take over, which can be costly and messy. If the termination occurs during the operation and maintenance (O & M) phase, compensation is likely. Fair compensation is necessary to attract bidders for the project, and it may be based on

- the actual cost incurred to date, plus a reasonable amount of profit;
- the remain asset value; or
- the present value of projected future profits.

Either way, the computation of the amount of compensation will be tedious and likely to be contentious unless the methods of compensation are written into the PPP contract. The method of computation will depend on the time of termination, among other factors. If termination is early, if, for example, the grantor has serious concerns over the operational performance of the SPV, then the cost approach is advantageous to the grantor because the compensation is likely to be lower. Further, there is likely to be objection to paying the SPV based on projected profits or work that has yet to be performed.

Project participants do not take termination lightly. If the SPV is facing financial difficulties, there may be other options, such as equity injections by existing or new investors, restructuring of loans, or renegotiation of the PPP contract. The latter includes lengthening the concession period, tariff adjustment, or changes in performance criteria.

Handing over issues

If there is no early termination, preparation for handing over the assets begins about three years before the end of the PPP contract. Depending on the PPP's contract terms, the issues concern the quality of the asset and possible compensation if the asset is of poor quality.

The grantor will carry out an early inspection and quality audit to determine which assets require improvement prior to handing over. The grantor may carry out the audit, or both parties can appoint an independent assessor.

If the quality of the assets is unsatisfactory, the SPV will have to make good or compensate the grantor. Since the handing over of the assets can occur decades after the project was conceptualized, it is possible that the asset may suffer from economic, physical, or technological obsolescence. In this case, the asset will need to be upgraded to newer standards rather than merely improved to satisfy the initial contract specifications.

Finally, the operator will also need to hand over the documents, including operating manuals, reports, drawings, records, logs, materials, and spare parts. Importantly, the operator has a long time series of operating data that are useful for predicting asset performance and energy consumption using evidence-based maintenance and data analytics, as discussed in the previous chapter. Hence, the quality and types of data to hand over should be a contract requirement.

By studying the performance, the grantor may appoint the same operator during the initial years to smoothen the transition, which includes training new operatives. Alternatively, the grantor may engage a new operator.

Public asset management

Public infrastructure asset management (Uddin *et al.*, 2013) is similar to private asset management, except that public value replaces private profitability. The principles are the same (see Chapter 14), such as

- the development of goals and objectives based on an assessment of needs and challenges;
- the formulation of a strategic asset management plan; and
- building organizational capacity.

In addition, there should be an enabling national and policy environment for public asset management. To ensure financial sustainability, some analysts have proposed the setting up of an urban wealth fund at arms-length from short-term political influence for the long-term investment and maintenance of infrastructure assets (Detter and Folster, 2017).

There is growing awareness of the need to improve the resilience of public infrastructure assets because of climate change (OECD, 2021). As discussed in Chapter 1, one can take a management or design/planning approach to develop resilient infrastructure systems.

References

Detter, D. and Folster, S. (2017) *The public wealth of cities: How to unlock hidden assets to boost growth and prosperity*. Washington DC: Brookings Institution.

OECD (2021) Building resilience: New strategies for strengthening infrastructure resilience and maintenance. *OECD Public Governance Policy Papers*, No. **5**. Paris: OECD Publishing.

Uddin, W., Hudson, R., and Haas, R. (2013) *Public infrastructure asset management*. New York: McGraw-Hill.

Index

accounts payable, 50
accounts receivable, 50
accrual principle, 133
adjudication, 211
agency CM, 110
agile project management, 22
American option, 107
arbitration, 211
Asian financial crisis, 7
asset enhancement initiatives, 226
asset management action plan, 220
asymmetric information, 14

Balance Sheet, 128
balloon loan, 122
bankability, 68
bankruptcy remoteness, 34
Belt and Road Initiative, 29
best-efforts deal, 154
best-value bids, 150
bid bond, 94
Bills of Quantities, 180
Black–Scholes model, 103
blue bond, 11
border prices, 46
bridges, 35
builder's all-risk insurance, 136
builder's risk insurance, 136
Building Information Modeling, 177
Building Wellness, 215

build-operate-own, 80
build-operate-transfer, 21
business cycles, 6

call option, 102
capacity charge, 35
capacity utilization, 197
Capital Asset Pricing Model, 119
capitalization rate, 75
carbon credits, 9
carbon emissions, 8
carbon regulation, 8
carbon tax, 8
carbon trading, 9
cash flow table, 128
catastrophe bonds, 137
central planning, 37
Certificate of Occupancy, 213
Certificate of Substantial Completion, 213
change order, 202
charges on movable assets, 157
Clawback clause, 159
Clean Development Mechanism, 9
climate bond, 11
climate change, 231
close-out activities, 213
club deal, 154
collared financing, 128
commercial risks, 64

commissioning plan, 178
commitment letter, 117
communications plan, 208
completion risks, 64
complex system, 13
conceptual design, 33
condition-based maintenance, 223
Construction Quality Assessment
    System (CONQUAS), 200
consequential loss, 136
constructability review, 172
construction tender documents, 179
constructive change, 202
contingent equity, 34
contingent valuation, 43, 52
conversion factor, 47
core inflation, 126
corrective maintenance, 222
corruption, 79
cost, insurance and freight, 206
cost overrun, 34
cost-effectiveness analysis, 40
counterfactual, 31
crashing, 194
credit enhancements, 163
credit rating agencies, 12
critical chain approach, 194
critical path, 192
cure period, 140
currency risks, 123
currency swap, 124

daily report, 209
deep learning, 225
default judgment, 211
default triggers, 155
defects, 205
defects liability period, 214
depreciation, 130

deregulation, 36
desalinated water, 35
Design Brief, 171
design development, 173
design proposal, 97
design review, 172
design work breakdown structure,
    172
Design-Build contract, 108
Design-Build-Finance-Operate, 35
Design-Build-Operate-Own, 35
Development Agreement, 95
development charge, 73
development program, 29
devolution, 14
differing site condition, 202
digital twin, 226
diminishing returns, 4
discount rate, 55
discourse, 81
disequilibrium models, 101
dispute resolution, 87
dispute review board, 211
documentation plan, 207
double-counting, 45
due diligence, 21
Dutch auction, 148

Early Contractor Involvement, 109
earned value, 198
EBITDA, 129
elemental cost approach, 111
elemental costing, 174
employer's liability insurance, 135
Engineering, Procurement, and
    Construction (EPC) contract, 109
English auction, 147
environmental impact study, 79
environmental liability insurance, 136

environmental, social, and governance (ESG) framework, 13
Equipment Contract, 144
equity, 116
equity internal rate of return, 132
European option, 107
export credit agencies, 117
extension of time, 211
externalities, 14

facilities management (FM) contract, 221
fee mechanism, 87
final payment, 213
financial risks, 64
fiscal sustainability, 78
fixed rate loan, 127
float, 193
floating rate loan, 127
force majeure, 22
force majeure risks, 64
forward purchase contract, 113
Fourier transform, 224
free cash flow, 131
free on board, 206
free-rider problem, 44
front-load, 185
frustration of contract, 211
Fuel Supply Agreement, 143
futures contract, 115

general liability insurance, 136
Geometric Brownian Movement, 104
geotechnical report, 170
global financial crisis, 7
Gordon's formula, 119
governance, 14
government grants, 117
government support agreement, 97

governmentality, 81
gray infrastructure, 10
green bond, 11
Green Building, 215
Green Climate Fund, 9
green infrastructure, 9
greenwashing, 11
guarantees, 34

hard costs, 45
hard facilities management services, 222
headline inflation, 126
hedonic pricing, 51
high-speed rail, 30
hold point, 207
hot money, 127
human capital approach, 51
hurdle rate, 59
hydroelectric power plant, 73
hydropower plant, 137

import-export banks, 12
income approach, 75
Income Statement, 128
index number approach, 196
infrared thermography, 225
infrastructure funds, 227
infrastructure groups, 227
infrastructure investment funds, 117
inspection plan, 207
insurance companies, 117
Integrated Project Delivery, 109
integrated project management, 26
interest rate swap, 128
internal rate of return, 57
inter-bank rate, 127
inter-dependency, 12
Invitation for Expressions of Interest, 85

in-principle approval, 153
Islamic finance, 117
Itô's lemma, 104

Japanese auction, 148

kaizen, 200
Kaldor–Hicks criterion, 54
Kanban, 23
Keynesian, 4
kick-off meeting, 188
Kondratieff wave, 7

labor market, 49
land acquisition, 91
Land Lease Agreement, 144
land titles, 73
leadership, 23
Leadership in Energy and
    Environmental Design, 215
learning by doing, 196
lenders' approval, 160
lender's clauses, 137
Letter of Intent, 151
Letters of Credit, 144
life cycle costing, 10
limited liability company, 166
limited recourse, 34
liquidated damages, 214
litigation, 211
loan covenants, 158
loan securitization, 162
long waves, 6
low-tax country, 166
lump-sum contract, 108

machine learning, 225
market feedback period, 93
market risk premium, 120

market risks, 64
market sounding, 82
market study, 99
master land use plan, 173
measurement contracts, 110
Monte Carlo simulation, 60
multilateral development agencies,
    117
Multilateral Investment Guarantee
    Agency, 118
multiplier effect, 45
municipal solid waste, 35

negative covenants, 160
net present value, 55
neural networks, 225
non-government organizations, 13
Notice to Proceed, 151
novating, 109
numeraire, 46

official exchange rate, 46
offshore escrow account, 124
off-balance sheet, 34
off-taker, 35
oil analysis, 224
operation and maintenance
    insurances, 137
operation and maintenance risks, 64
opportunism, 67
optimal capital structure, 118
option pricing, 103
over-bidding, 149

parametric method, 111
Pareto criterion, 54
Paris Agreement, 9
Partial Risk Guarantee, 158
partnering agreement, 189

pass-through mechanism, 78
pay when paid approach, 207
payment models, 35
penalties for poor performance, 143
pension funds, 117
performance bond, 113
performance obligations, 73
Plan-Do-Check-Act cycle, 200
plot ratio, 76
points system, 86
political risks, 64
possible assignment of benefits, 87
predictive maintenance, 223
preventive maintenance, 222
pre-feasibility study, 32
price elasticity of demand, 39
price proposal, 98
prime cost sum, 180
privatization, 36
procurement strategies, 16
productivity, 196
productivity change method, 43
professional liability insurance, 136
programming, 170
progress report, 208
project baseline plans, 185
Project Brief, 71, 167
project buffer, 194
Project Charter, 71
project delivery models, 16
project governance structure, 15
Project Initiation Document, 71
project integration management, 26
project management office, 166
project management plan, 186
project options, 61
project scope, 32
prospect theory, 52
provisional sum, 180

public infrastructure asset
    management, 231
punch lists, 213
put option, 103

Quality Assurance, 198
Quality Control, 198

railways, 35
random walk, 104
ratio analysis, 134
real estate investment trusts, 6
real options, 63
refinance, 162
regulatory risks, 64
reinsures, 137
reliability-centered maintenance,
    222
Request for Information, 86
Request for Proposal, 21, 85
Request for Qualification, 85
resilient infrastructure, 11
resource leveling, 194
Response Package, 85
retention sum, 112
revealed preference approach, 51
revenue equivalence theorem, 149
revocation of contract, 96
risk premium, 56
risk-adjusted discount rate, 61

scale economies, 196
Schedule of Basic Rates, 179
schedule of values, 185
schematic design, 173
school, 73
scope creep, 201
scrap values, 47
sealed bid auction, 148

securitization, 6
security deposit, 112
Security of Payment Act, 207
sensitivity analysis, 60
shadow bid model, 77
shadow exchange rate, 46
shadow prices, 43
Shareholders' Agreement, 95
site analysis, 73
site appraisal, 74
social and environmental risks, 64
social capital, 2
soft costs, 45
soft facilities management services, 222
soft skills, 25
sovereign wealth funds, 4
space program, 172
speculation, 125
speculative building, 101
stakeholder management, 203
Standard Brownian Movement, 104
standard forms of contracts, 141
start-up and testing, 213
Statement of Work, 71
statistical value of life, 44
step-in rights, 87
stock-flow model, 99
structural reforms, 36
submittals, 197
subordinated debt, 155
subprime mortgage, 7
summary judgment, 211
surrogate market, 51
sustainable development, 79
syndicated loan, 161

take-or-pay purchase contracts, 102
tender deposit, 94
Through-put Agreement, 143
toll roads, 35
Tolling Agreement, 143
Tornqvist index, 196
Total Quality Management, 198
traffic demand, 41
transfer payments, 45
trust, 2

ultrasound technology, 225
unincorporated business entity, 166
unsolicited proposals, 29
user pays, 35

value engineering, 172
value for money, 36
variation order, 202
vibration analysis, 224
Vickrey auction, 148

warranties, 112
waste-to-energy, 35
waterfall cash flow model, 161
waterfall project cycle, 22
weighted average cost of capital, 56
Wiener process, 104
willingness to accept, 52
willingness to pay, 38
winner's curse, 149
with and without, 31
Work Breakdown Structure, 168
work injury compensation insurance, 135
working capital, 49